T0406493

Environmental Science

Series Editors: R Allan • U. Förstner • W. Salomons

M.C. Th. Scholten, E.M. Foekema,
H.P. Van Dokkum, N.H.B.M. Kaag and
R.G. Jak

Eutrophication Management and Ecotoxicology

With 75 Figures

Springer

Martin C.Th. Scholten
Edwin M. Foekema
Henno P. Van Dokkum
Nicolaas H.B.M. Kaag
Robbert G. Jak
TNO-MEP
P.O. Box 57
1780 AB, Den Helder

The Netherlands

E-mail:
H.vanDokkum@mep.tno.nl

ISSN 1431-6250
ISBN 3-540-22210-3 **Springer Berlin Heidelberg New York**

Library of Congress Control Number: 2004114232

Springer is a part of Springer Science+Business Media
springeronline.com

Printed in Germany

Cover design: E. Kirchner, Heidelberg
Production: A. Oelschläger
Typesetting: Camera-ready by PTP-Berlin
Printing: Mercedes-Druck, Berlin
Binding: Stein + Lehmann, Berlin

Printed on acid-free paper 30/2132/AO 5 4 3 2 1 0

Table of Contents

1 Eutrophication and the Ecosystem 1
1.1 Background 1
1.2 Eutrophic and Eutrophicated Waters 2
1.3 Trophic Cascades in Freshwaters 5
1.4 Eutrophicated Waters: Disturbed Trophic Cascades 13

2 Daphnid Grazing Ecology 21
2.1 Daphnids and Their Ecological Role 21
2.2 Daphnid Grazing Effectiveness – Concept and Measurement 38
2.3 Modelling Daphnid Grazing Effectiveness 49

3 Toxic Reduction of Daphnid Grazing Effectiveness 57
3.1 Daphnid Ecotoxicology 57
3.2 Observations from Mesocosm Studies 62
3.3 Daphnid Grazing Effectiveness in Response to Toxicant Exposure 72

4 Field Observations of Daphnid Grazing 81
4.1 Two Different Lakes in Holland 81
4.2 The Plankton Dynamics in Lake Geestmerambacht 82
4.3 The Plankton Dynamics in Lake Amstelmeer 89
4.4 What Can Be Learned from These Lakes? 91

5 New Perspectives for Eutrophication Management 97
5.1 A New Dimension in Lake Eutrophication Management 97
5.2 How to Address Ecotoxicology in Eutrophication Management 102
5.3 Epilogue 108

Acknowledgements 111

References 113

1 Eutrophication and the Ecosystem

1.1 Background

Eutrophication of surface waters is generally recognised as a matter for environmental concern. Eutrophication is characterised by increased algal growth, with an increased incidence of toxic cyanobacteria blooms and a decrease in the abundance of species.

Some of the manifest problems brought about by prolific algal biomass include: turbid waters; anoxic conditions; bad smell and chironomid and Culex midge plagues (Vollenweider 1990; Moss et al. 1996a; Carpenter et al. 1998). Such eutrophication problems ("eutrophication" *sensu lato*) are generally considered to be the consequence of enhanced nutrient loadings ("eutrophication" *sensu stricto*) (Likens 1972; Vollenweider 1990; Reynolds 1992; Moss et al. 1996a; Carpenter et al. 1998). Therefore, the management of eutrophicated water bodies is usually primarily focused on the reduction of nutrient loading, supported by a policy of reduced environmental releases of phosphorus from laundry detergents, sewage and agriculture.

However, it became apparent over the past decade, that reduced grazing of algae by daphnids can be a crucial factor determining whether or not nutrient enrichment will lead to eutrophication problems (Moss et al. 1991; Moss et al. 1996b; Reynolds 1994). Biomanipulation of eutrophicated shallow water bodies, thereby improving ecological conditions for daphnids, became a regular tool applied in eutrophication management practice (Benndorf 1990; McQueen 1998; Harper et al. 1999).

Biomanipulation is mainly focussed on the improvement of biological conditions leading to a higher survival rate for daphnids as part of the aquatic foodweb. Examples of biomanipulation measures include: reduction of predation by planktivorous fish and improvement to the submerged vegetation as a shelter for daphnids against predation. More recently, the palatability of suspended particles as a factor determining the grazing efficiency of daphnids has become a topic of interest. High concentrations of resuspended inorganic particles hamper daphnid grazing, while at the same time stimulating algal growth due to increased nutrient releases (Kirk and Gilbert 1990; Ogilvie and Mitchell 1998). Top-down control by daphnids under eutrophicated conditions may also be reduced by the presence of unpalatable algal species (mainly cyanobacteria) that may gain competitive advantage over the heavily grazed palatable algal species.

This book covers another important aspect regarding the improvement of environmental conditions for daphnids, which is necessary for successful eutrophication management, i.e. optimalisation of the abiotic water conditions. Bales et al.

(1993) suggested that the sensitivity of daphnids to saline conditions may be a significant reason for the higher susceptibility of brackish waters to eutrophication. From this perspective, ecotoxicologically reduced daphnid grazing due to micropollutant loadings could be a crucial factor leading to problems associated with eutrophication (Hurlbert et al. 1972; Hurlbert, 1975; Gliwicz and Sieniawska, 1986).

The toxic effects of pesticides and other chemicals on the viability of cladoceran populations reduce their capacity to graze the surplus algal growth caused by increased nutrient availability. In ecotoxicological semi-field studies, it has been observed that cladocerans are amongst the most sensitive species when it comes to toxicant exposure, consequently resulting in a reduction in the top-down control of the algal growth (Day 1989; Yasuno et al. 1993). Based upon the analysis of sediment cores, Stansfield et al. (1989) argued that a switch from submerged plant dominance to phytoplankton dominance (eutrophication) in a series of shallow lakes, i.e. the Norfolk Broads, U.K, during the 1950s and 1960s, was likely to have been due to the poisoning of cladocerans (viz. *Daphnia*) as a result of liberal organochlorine pesticide use.

The aim of this book is to provide a better understanding of the ecotoxicological aspects of eutrophication processes in shallow, temperate fresh waters, so that these processes may become a recognized factor in the restoration of eutrophicated water bodies.

Some basic limnological ecology, which is fundamental to the further contents of this book, is presented in the first chapter. Chapter 2 provides more information on daphnids, encompassing their ecology, grazing efficiency and any subsequent consequences for the control of algal densities. Chapter 3 concerns daphnid ecotoxicology, and provides information on toxicity induced reduction of daphnid grazing effectiveness (so called "toxic anorexia") in experimental settings. Variation in daphnid grazing effectiveness in the field situation is described for two Dutch lakes in Chap. 4. The applicability of an ecotoxicological assessment of eutrophicated water bodies is discussed, and practical tips given, in Chap. 5.

1.2 Eutrophic and Eutrophicated Waters

Nutrients or Algae

In order to acquire a better understanding of the causes of eutrophication in fresh water ecosystems, it is helpful to make a distinction between **"eutrophic" waters** (classification of water according to its intrinsic nutrient status, eutrophication *sensu stricto*) and **"eutrophicated"** (or "eutrophied") **waters** (perception of manifest water quality problems related to ecological malfunctioning such as turbid water, bad smell and high algal density, eutrophication *sensu lato*).

The total phosphorus concentration of fresh surface waters is generally used as an indicator for the trophic status of that water body. Phosphorus is one of the essential nutrients for algal growth. It is considered to be the prime limiting element determining the biological productivity (algal productivity, and subsequent higher order productivity) in many freshwater aquatic systems.

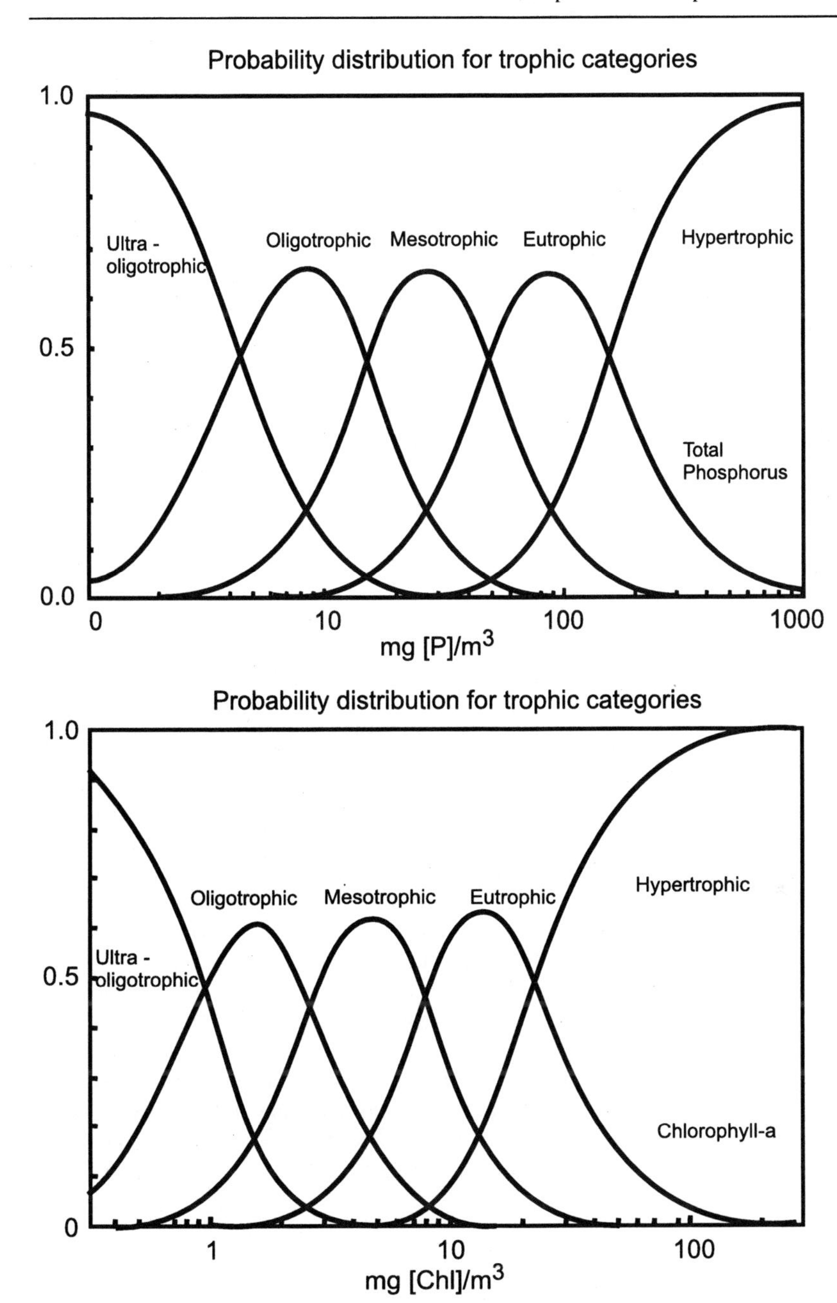

Fig. 1.1. Classification of waters according to the OECD (1982). Above: on the basis of P-loading, and below: on the basis of average algal density

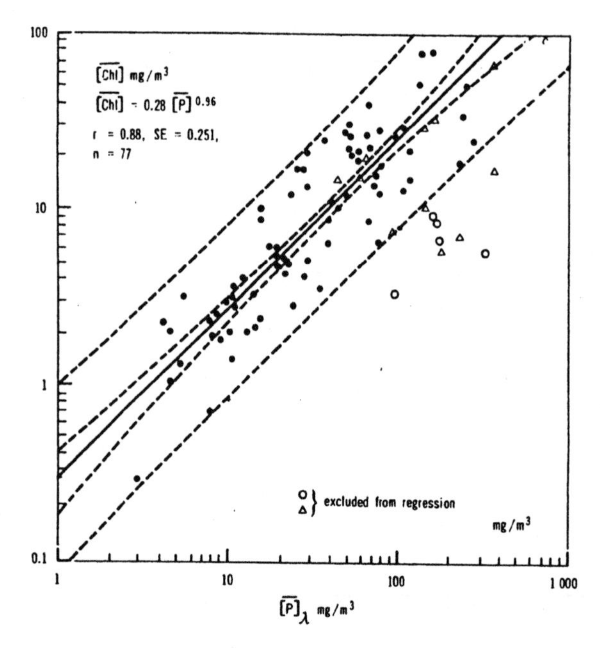

Fig. 1.2. The "Vollenweider-model" (OECD 1982) illustrating large variation in the P:chl-a ratio

Other main nutrient elements present in potential limiting concentrations are nitrogen and silicate. Nitrogen is not a suitable indicator, because it is poorly retained by soils and leaches easily to aquatic systems. Additionally, transformations between various fractions of particulate (organic) nitrogen, dissolved (organic/inorganic) nitrogen and atmospheric (inorganic) nitrogen occur at high rates, ensuring a continuous supply of available nitrogen under most circumstances.

Silicate (ortho-silicic acid) is only essential for the growth of diatoms, which incorporate silicic acid into their frustulers. Therefore, it is not a suitable indicator for total system productivity.

In the classification of the trophic state of waters according to the OECD (1982), no distinction was made between eutrophic and eutrophicated waters. Waters are referred to as mesotrophic, eutrophic or hypertrophic on the basis of their P-loading as well as on the basis of algal densities (Fig. 1.1), with the underlying assumption that nutrient status (P loading) is positively and causally correlated with algal density (Fig. 1.2).

The Vollenweider Regression

The OECD classification was based upon a regression model (known as the "Vollenweider" model) in which the phosphorus concentration (as indicator for the trophic status) of surface waters and mean algal density (as indicator for manifest "eutrophication" problems) were related to each other. The regression model was based upon observed relationships between phosphorus and algal densities in various deep upland lakes during the nineteen seventies (Vollenweider and Kerekes 1980). The model has been used often and validated in many situations (CUWVO 1976, 1980, 1988; OECD 1982; Klein 1989). This has led to the perception that the ecological water quality problems related to high algal densities were solely due to "eutrophication" (i.e., nutritional or, more specifically, phosphoric loading of waters).

It should, however, be noted that the "model" demonstrates a broad variation in the relationship between the two indicator parameters (a factor of 10 at 99% confidence), even after plotting on a log-log scale (Fig 1.2). This means that many eutrophic waters exist with relatively low algal densities; and oligotrophic/ mesotrophic waters with relatively high algal densities. Additionally, it should be noted that examples of eutrophic lakes with a relatively low algal density were excluded from the regression (open symbols in Fig 1.2).

Beyond Nutrients

Although phosphorus is a suitable rough indicator for the trophic status of a water body, this does not imply that it is a reliable (causal) predictor of manifest eutrophication problems (Reynolds 1992). The relationship between phosphorus content and eutrophication problems is not fixed, due to the fact that there are many other variables, in addition to phosphorus concentrations, that influence the manifestation of eutrophication problems. In certain circumstances, eutrophic waters are not eutrophicated, while mesotrophic waters sometimes appear to be eutrophicated.

The "Vollenweider-model" had its function in clarifying the role of excessive phosphorus loads in the manifestation of severe eutrophication problems, but for successful eutrophication management it is necessary to look beyond nutrient loads in order to arrive at reliable and sustainable water management options (Golterman 1991; Moss et al. 1994). Nutrient-algal relationships are only a fraction of the complete aquatic food web (Hosper et al. 1992; Scheffer et al. 1993; Leibold and Wilbur 1992; Vijverberg et al. 1993). An extensive statistical data analysis of 231 lakes in the Netherlands (Portielje and van de Molen 1997a; 1997b; 1998) has made it clear that top-down control of algal density development by submerged vegetation (competition) and daphnids (grazing) is a significant factor explaining variation in chlorophyll nutrient ratios. In eutrophication management it is important to understand what has caused the eutrophicated state of the waters, and to recognise the bottleneck situations inhibiting their restoration.

1.3 Trophic Cascades in Freshwaters

The Foodweb

The availability of nutrients is a basic requirement for the development of organic biomass. Algae can respond quickly and opportunistically to increases in nutrient availability. However, the capacity of individual algae to store nutrients is limited. It is the population that retains the nutrients, but only during the growing season.

Decaying algae release nutrients and the following season all nutrients must be re-acquired.

Macrophytes on the other hand, cannot respond as quickly as algae to changing nutrient levels, but do retain nutrients individually within storage organs. Macrophytes are also able to acquire nutrients, both actively and passively, from the substrate on which they grow. This gives them a competitive advantage over algae. The shade created by macrophyte leaves may inhibit algal growth. Some macrophytes can even suppress algal growth by releasing allelochemicals (Mjelde and Faafeng 1997). By preventing algal growth, macrophytes may also gain access to nutrients in the water that might otherwise be utilised by the more rapidly responding algae. As a result, macrophytes are the predominant primary producers in shallow waters where sunlight can reach through to the sediment, whereas algae are the primary producers in deeper waters.

The primary production of algae and macrophytes is the basis of aquatic food webs (see Fig 1.3). The main groups of algal consuming secondary producers in fresh water ecosystems are cladocerans and other zooplankton groups (viz. copepods and rotifers) and filter feeding benthic species (viz. bivalves). Zooplankton (especially cladoceran water fleas) plays a key role in aquatic ecosystems by effectively responding to variations in algal production through its opportunistic population dynamics. Bivalves, though individually capable of filtering large volumes of water, do not play such a central role, since they cannot multiply at the same rate as their algal food source. Therefore, bivalve populations are limited by the minimum amount of food available.

The algal biomass density is the nett result of algal production and algal losses due to sedimentation and grazing by zooplankton or other secondary producers. The production of herbivore biomass in aquatic ecosystems is approximately 2–6% of the primary (i.e., algal) production, which is extremely high compared to terrestrial ecosystems (<< 1%) (Barnes and Mann 1993). Accounting for respiration and defaecation losses, this means that in a healthy aquatic ecosystem that 30–90% of the primary production is consumed and ends up in the aquatic food chain. In plankton dominated systems this is at the higher end of the scale (up to 90–95%). Only a small fraction of the primary production is allocated for the establishment of a standing vegetation (e.g., reed and rush marshes) and temporary algal blooms. Sedimentation and shore deposition of unconsumed algal and plant debris (detritus) brings about decay by benthic detrivores.

Zooplankton is consumed by small fish (e.g., roach and smelt, and the juveniles of larger species) and amphibians, which, in turn, are food for predatory fish (e.g., pike and perch). Zooplankton predation, especially of the larger cladoceran species, may be very high. Therefore, cladocerans usually seek refuge in the macrophyte vegetation and migrate out into open water only during the night when the risk of predation is low (Timms and Moss 1984; Lauridsen and Lodge 1996; Stephen et al. 1998). Zooplanktivorous fish, on the other hand, avoid macrophyte vegetation due to the fact that predators such as pike may be concealed there.

The aquatic food chain provides a food source for many birds (e.g., waterfowl) and various mammals (e.g., otters).

The Role of Macrophytes

Macrophytes as seen in figure can play substantial role in shallow lakes and ponds as they contribute to primary production of organic material that give organic matter to the bacterial loop and invertebrates. Of course, in high trophic and turbid lakes, this process do not occur as a significant contribution in nutrient cycles as lack of light is limitant factor for their growth.

We must distinguish the different types of macrophytes as shown in the (following) figure:

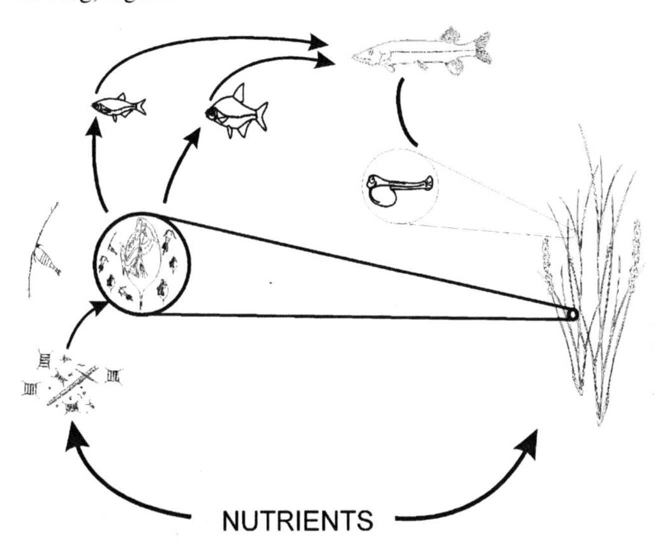

Fig. 1.3. The position of zooplankton (i.e. cladocerans) in the foodweb

As any plants, macrophytes participate to the productivity of the ecosystem. When rooted macrophytes are present, they can export nutrient from sediments and at their decay and senescence contribute to the enrichment of the water column. This phenomenon may be temporarily decreased by periphyton fixed on leaves and stems, nevertheless, N and P are rapidly available for small algae and heterotrophic microflora. Senescing macrophyte tissue and attached algae, as well as organic matter and nutrients retained in them stay in sediments where they collapse in the littoral zone of lakes. The accumulative retention of these inputs of nutrients and carbon can be very high. It should be recalled that most aquatic macrophytes are perennials and much of certain nutrients as phosphorus extracted by plant growth from the sediments is largely translocated back to the rooting tissues at the end of the growing season and taken off from the nutrient pelagic cycle of lakes. In case of shallow lakes and high development of littoral zone, the total nutrient releases to the pelagic zone is reduced. The tendency is for the littoral developments and metabolism to reduce total phytoplankton growth of the lake to levels below the growth potential that would occur if most of the nutrient loads entered the pelagic zone directly. Wetzel (1990) has shown that in these conditions, "in the ecosystem with profundal sedimentation patterns and the converse differences in system components and their metabolism, one finds a overall decrease in the phytoplanktonic efficiency of nutrient utilization". The relations between P and macrophytes have been discussed also for a long time and it is known that these plants are able to keep quite large quantities as "luxury consumption". In summer, this contribution of macrophytes in the P cycle can participate of decreasing concentrations of the growth capacity of algae in the littoral zones of the lakes, but in autumn, large amounts of P are released. In the same way, the allocation of macrophytes in toxicants explains the leaching of these compounds under specific conditions (pH, redOx...).

Balanced Foodchains

In a healthy ecosystem, a finely tuned balance exists between production and consumption within each link in the food chain (Carpenter et al. 1985). In shallow freshwater bodies, macrophytes are important for the stabilisation of this balance. Their long-term accumulation of nutrients tempers phytoplankton dynamics and at the same time they offer shelter for zooplankton, protecting it from predation. To a certain extent, bivalves also have a stabilising function, as they graze a base level of algae and clear the water of any suspended particles that may interfere with zooplankton grazing.

However, zooplankton, especially cladocerans, controls the short-term variation in phytoplankton dynamics, through their opportunistic response to food availability. The opportunistic response of cladocerans to algal (food) conditions is thus a critical and central step in the trophic cascade of aquatic foodchains. A well-balanced and synchronised coupling of cladoceran development to algal development is, required therefore, especially under mesotrophic or eutrophic conditions.

The Trophic Basis of the Cascade

Nutrients for algal growth become available through the mineralisation of organic materials, organism excretions and leaching from sediments. The cycles of the macronutrients C, N and P are the best understood and are briefly described below. Trace elements may limit primary or secondary production both spatially and temporarily. This may, for instance, be the case for Si, which is an important element for diatoms (Hecky and Kilham 1988; Moss and Balls 1989). The cycles and importance of trace elements are less well understood and will, therefore, not be further considered here.

The *carbon cycle* is the basis of all biogenic processes. Atmospheric CO_2 is fixed by autotrophs and enters the food chain, where a major portion of it is rereleased by means of heterotrophic metabolism. Heterotrophic consumers rely, both directly and indirectly, on this autotrophic fixation of CO_2 for their carbon supply. Carbon is, therefore, often used to quantify energy fluxes in ecosystems. Generally, carbon is not considered to be a limiting nutrient for primary production, due to the large available pool of atmospheric CO_2. In the aquatic environment, however, C may become temporarily limiting, especially during summer stratification, when the rate of autotrophic CO_2 fixation exceeds the combined fluxes of heterotrophic CO_2 production and the transport of atmospheric CO_2 into the water.

The *nitrogen cycle* is very dynamic and is characterised by many transformations, in which (micro)biological processes play an important role. Nitrogen primarily enters the biogenic cycles by microbial nitrogen fixation of atmospheric N_2 (Vitousek et al. 1997; Galloway 1998; Pastor and Binkley 1998). Nitrogen compounds are not absorbed by soil, clay or ferric hydroxide, etc. A large fraction of the N, therefore, dissolves in either inorganic (ammonia, nitrate, nitrite) or organic (amino acids, proteins, nucleotides) forms and easily leaches to surface waters (Overbeck 1989). Part of the N is lost to the atmosphere due to denitrification. In contrast to terrestrial ecosystems, which are often characterised by N limitation, most freshwater systems have a sufficient supply of N from terrestrial run off. Within aquatic systems, the N cycle is primarily controlled by microbial processes (nitrification, denitrification, ammonification), which are strongly dependent on the redox status of the system (Overbeck 1989; Stumm and Morgan 1996).

The *phosphorus cycle* is mainly driven by physico-chemical processes. P readily absorbs to soil and, thus, most P is particle bound and only a fraction is directly available to biota (Sharpley and Rekolainen 1997). The most important natural route from terrestrial to aquatic ecosystems is terrestrial run off and erosion of particulate P (Sharpley et al. 1995). The bulk of the P in water is in particulate form and much P is bound to the sediment. The bioavailable fraction, the soluble orthophosphate (SRP), is very small and has an extremely high turnover rate (minutes). The P cycle in lakes is heavily dependant on the redox conditions of the system. Under anoxic conditions, the soluble P fraction drastically increases and substantial amounts of P may be released from the sediment. Under oxygenated conditions, soluble inorganic P is readily bound by adsorption to ferric hydroxide and $CaCO_3$ and by precipitation as ferric phosphate. The oxygenated surface layer of the sediment acts as an efficient P trap (Overbeck 1989; Grobbelaar and House 1995). Under certain conditions, P bound to sediment may be utilised by phytoplankton (Golterman 1977; Grobbelaar 1983; Grobbelaar and House 1995). However, this is negligible in clear shallow water due to the presence of a layer of benthic diatoms, which stimulate an oxygenated sediment surface layer and arrest nutrient fluxes from deeper sediment layers. The remaining P fraction is used by benthic diatoms for growth (Van Luijn et al. 1995). Additionally, the layer of benthic diatoms stabilises the sediment surface, thereby reducing the amount of resuspended sediment particles. Bioturbation by sediment-dwelling organisms may increase P-fluxes from the sediment, but this effect is counteracted by the increased oxygenation of the surface layer of the sediment (Andersson et al. 1988). Assimilation in the biomass (e.g., submersed and shoreline vegetation, aquatic food chain) is an important factor in the P-cycle.

Hydrological Aspects

Resuspension of inorganic sediment particles is mainly caused by wave action and, occasionally, by foraging fish. These resuspended particles can act as an internal source of phosphate (Ogilvie and Mitchell 1998). The dynamics of nutrients is also determined by the depth and stratification of lakes. In deep lakes, thermal stratification occurs during winter and summer. Nutrients present in the epilimnion (upper water layer) will be transported to the hypolimnion (bottom water layer). The epilimnion of a stratified lake can lose up to 50% of its total phosphorus during the summer (Scheffer 1998). The 'turn over' in autumn and spring makes the nutrients from mineralised material in the hypolimnion available to the epilimnion. There is no stratification in shallow lakes and there is a continuous exchange of material between sediment and water. The mineralisation rate of sediment and the subsequent release of nutrients will increase due to higher temperatures in summer (Jeppesen et al. 1997).

Other environmental factors such as flushing rate, water temperature, pH and water hardness also influence growth rates and the composition of algal communities (Reynolds 1989; Moss and Balls 1989; Moss et al. 1991; Beklioglu and Moss 1995). However, since nutrients form the basis of ecosystem production, nutrient availability is the key factor affecting the functional role of algae as basic producers in the food web.

Nutrients and Algal Growth

The relative availability of different nutrients changes over time, as the nutrient cycles are not synchronised. In order to assess the relative availability of nutrients, the concentrations of macronutrients may be compared with the molar Redfield ratio of 106C:16N:1P (Goldman 1979). As only one growth factor can be limiting at a time (Liebig's Law of the Minimum), the Redfield ratio is considered to be an optimum ratio for algal growth, at which transitions from one nutrient limitation to another occur (Grobbelaar and House 1995). The Redfield ratio is an average ratio of algal elemental composition in marine waters, which is relatively stable compared to the nutrient concentrations in fresh waters. Such a stable elemental composition is thought to have evolved because algal cells perform similar metabolic functions and have quantitatively similar structural requirements (Hecky and Kilham 1988).

It should, however, be realised that different algal species have different optimum nutrient requirements, resulting in changes in community structure with changes in nutrient ratios (Grobbelaar and House 1995). This is especially true for fresh water systems, where algal elemental composition seems to be more variable, with respect to the Redfield ratio, than in marine systems (Hecky et al. 1993).

Since the relative availability of nutrients is not constant, but varies over time and between water bodies, most aquatic systems are resource limited. N and P are often the primary limiting nutrients. For our mesocosm studies (see Chap. 3), it has been observed that up to a depth of 80 cm, the P-flux from the sediment is sufficient to sustain algal growth and that the remineralisation of N, which is proportional to the height of the water column, is the limiting factor for optimal algal growth conditions. However, at a water depth of 180 cm, the P-flux, which was primarily related to the sediment surface area, was not sufficient to sustain algal growth. In lakes, this deeper water is the exclusive domain of algae.

To ensure survival, a competitor must be able to maintain nett population growth at resource levels less than those required by other species (Tilman et al. 1982). Algae are particularly well adapted for scavenging their environments for resources. Concurrent with the observation that P is most often the nutrient structurally limiting in their environment, these strategies are particularly suited to compensating for P shortage.

In response to nutrient limitation, algae may react with three different strategies, or a combination of these (Sommer 1989):

- affinity-strategists can utilise low nutrient concentrations very efficiently;
- growth-strategists utilise transient periods with high nutrient concentrations to achieve a rapid population growth, compensating for periods with low concentrations and slow growth;
- storage-strategists build up internal reserves of nutrients during periods of high nutrient concentrations (luxury uptake), which are utilised during transient periods with low nutrient concentrations.

Affinity strategists are not only very efficient in the uptake of scarce nutrients, they also release substances that increase the availability of bound P (Grobbelaar 1983; Grobbelaar and House 1995). The cyanobacterium *Anabaena,* for instance, excretes extracellular phosphatases almost immediately upon the onset of P limitation (Healy 1973). Algae may also change the pH of their surroundings, which also increases the availability of adsorbed P (Grobbelaar 1983). Flagellates may be seen as behavioural affinity strategists, since due to their motility, they can utilise patches of any available nutrients caused by zooplankton excretion (Sommer 1989).

Storage strategists may have internal P reserves that may be sufficient for 5 to 10 cell divisions. The storage capacity for other elements, on the other hand, is substantially lower; ca. 5 times lower for N and negligible for Si, even in the case of diatoms (Fisher et al. 1995).

Nutrients and Algal Communities

Varying nutrient concentrations, in an absolute as well as in a relative sense, will result in the development of different algal communities, depending on the outcome of intraspecific competition (Sommer 1989; Valiela 1993). The rate and quality of algal production is heavily dependent on the availability of both phosphate and nitrogen, and the subsequent N:P ratios. An analysis of phytoplankton data in European lakes (Schreurs 1992) shows that cyanobacteria dominate lakes with relatively low fractions of soluble reactive phosphorous (SRP), while green algae dominate systems with higher SRP (Fig. 1.4). Blue-green dominance increases with total N concentrations (Fig. 1.5). On the basis of total P, cyanobacteria dominate moderate classes (100–800 $mg.m^{-3}$), while green algae dominate at higher levels (>800 $mg.m^{-3}$). Lakes with low nutrient concentrations encompass a significant representation of flagellates in their phytoplankton communities.

The observations of blue-green dominance at high N:P ratio is contrary to what is generally assumed and observed in chemostate experiments (Andersen 1997), where cyanobacteria prefer a low N:P ratio. The dominance of cyanobacteria in fresh waters is, on the basis of these observations, often attributed to their ability to fix N by means of heterocysts. As N-limitation can be avoided by N-fixation, it has been hypothesised that the P-demand of blue-greens is high and that cyanobacteria may therefore have an advantage at low N:P ratios. On the contrary, however, it can be argued that high N:P ratios favour blue-green algae, since eco-

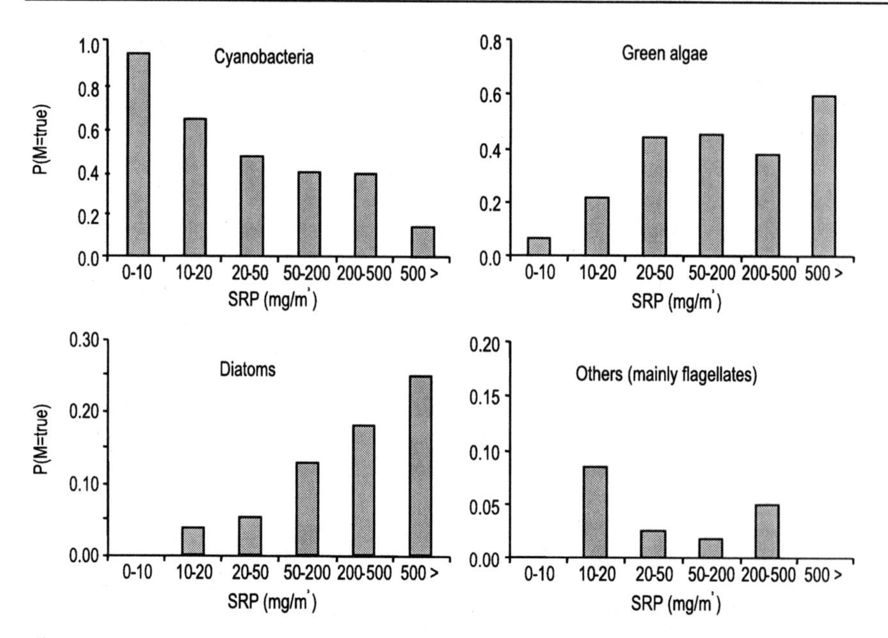

Fig. 1.4. Data from European lakes show that cyanobacteria have a lower frequency of dominance in P-rich waters, which favour green algae and diatoms (from Schreurs 1992)

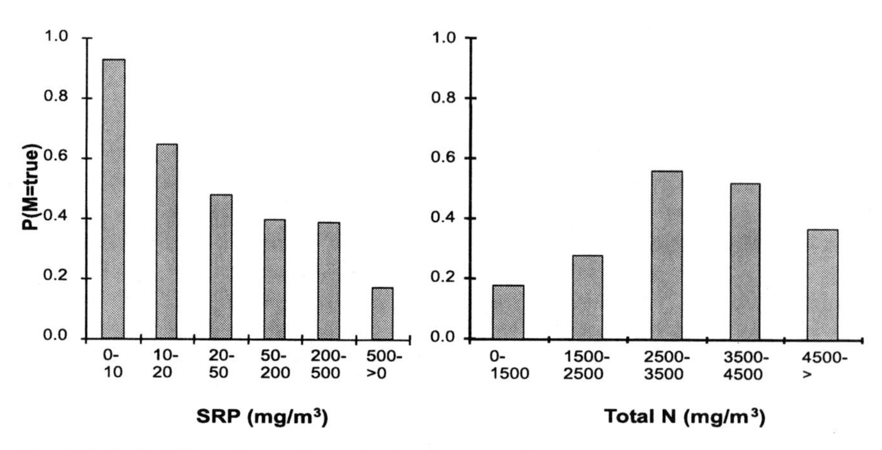

Fig. 1.5. Probability of cyanobacteria dominance in freshwaters at various P or N conditions (from Schreurs 1992)

physiological studies with these algae have demonstrated their more efficient use of P in comparison to other algal taxa and/or their better storage abilities. This high affinity for phosphorous results in a high nitrogen demand, and this may be the reason for the evaluation of facilities for cyanobacteria to fix nitrogen gas. From this point of view, a high availability of N in relation to P may increase the incidence of nitrogen fixing cyanobacterial blooms.

Although nutrients are often limiting for algal growth, other factors might be just as important (Reynolds 1989). Light, used as the energy source for obligatory autotrophic algae, is often available at low intensities due to high water turbidity or deep mixing of the water column (Bleiker and Schanz 1997). In water containing high levels of suspended matter or algal cell concentrations, light is scattered by particles and absorbed by algal pigments and dissolved organic matter, and it may become the limiting factor.

Nutrients and Zooplankton

Planktonic algae are heavily grazed by zooplankton. Cladocerans are the most efficient grazers, due to their relatively large size and indiscriminate feeding habit. Because of differences in elemental stoichiometry, different types of zooplankton have a markedly different influence on the nutrient cycle. Relative to their food (algae), zooplankton species have low C:N and C:P ratios, which means that the caloric value of the food is generally sufficient, but that the macronutrients N or P may be limiting for growth. Copepods have relatively high N:P ratios (40–50:1) and, consequently, N is often the limiting nutrient in their diet (Hessen 1997). Copepods, therefore, preferentially select N-rich algal cells (Butler et al. 1989). Copepod excretion is fluid and does not sedimentate. Non-assimilated nutrients are, therefore, instantly replaced to the benefit of algal growth (Lyche et al. 1996a).

Cladocerans typically have a low N:P ratio of ca. 12–15 (Hessen 1997). Although they do not actively select P-rich food particles, cladocerans retain P from their food very efficiently (Lyche et al. 1996a). Cladocerans have solid faecal pellets that easily sedimentate and thus attribute to a nutrient loss from the epilimnion. Sedimentation of (dead) animals may reinforce P-limited conditions (Urabe et al. 1995; Rothaupt 1997). In this way daphnids create P-limited conditions in which they have a competitive advantage over other zooplankton, due to their efficient P-scavenging.

Cladocerans are the preferred prey of zooplanktivorous fish, resulting in increased P recycling within the system (Carpenter et al. 1992; 1995b; Vanni and Layne 1997). Predatory fish, on the other hand, reduce the number of zooplanktivorous fish, thereby releasing predatory pressure on cladocerans and retaining P in the biomass (Carpenter et al. 1995a).

1.4 Eutrophicated Waters: Disturbed Trophic Cascades

Eutrophication

Eutrophication problems are related to excessive algal densities. These problems arise due to prolific algal growth caused by nutrients becoming available to the algae in excess of the ecosystem's capacity to stabilise their concentrations in combination with the inability of its zooplankton to control algal development.

Manifest eutrophication problems are, thus, related to a complex of factors: an ecological imbalance between nutrient loading or mobilisation on the one hand, and herbivorous algal grazing and the development of water plants on the other. In "eutrophicated" water bodies, most of the algal biomass remains ungrazed (resulting in algal blooms) and ends up in the decomposition pathway (eventually resulting in oxygen depletion).

From the 1950s on, the nutrient load in lakes started to increase because of the rapid increase in the human population, and agricultural intensification accompanied by intense use of fertilizers. Other perturbations, such as chemical pollution (e.g., pesticides) toxic to zooplankton and the loss of lake-marginal wetlands, have also contributed to the effects of nutrient enrichment (Hosper 1997). These continuous and incidental sources led to higher algal production, eventually resulting in problems associated with eutrophication. Until recently, increased nutrient loads were considered to be the main source for the detoriation of the lake ecosystems. Eutrophication resulted in turbidity of the water and the increased production of algae reduced the light available tothe macrophytes which, therefore, disappeared. The elevated production of algae and the decaying macrophytes produced large amounts of detritus, which accumulated on the bottom of the lakes to form loose sediment. Resuspension of this loose sediment by wind, waves and benthivorous fish further enhanced the turbidity of the water (Meijer 2000).

Although the increased nutrient loads in themselves are necessary prerequisites to prolific algal development, attempts to restore water bodies showed that nutrient reduction alone is not sufficient, unless extensive reductions are achieved. A survey of the trophic status of 231 Dutch lakes, covering the period 1980–1996, indicated a significant reduction in nutrient levels and chlorophyll concentration in nearly half of the lakes (Portielje and Van der Molen 1998). However, a more close evaluation of water quality data from the Netherlands shows that this is mainly due to reduced peak levels and that the reduction in chlorophyll concentrations is much less pronounced than the reduction in the P-loads (Fig. 1.6).

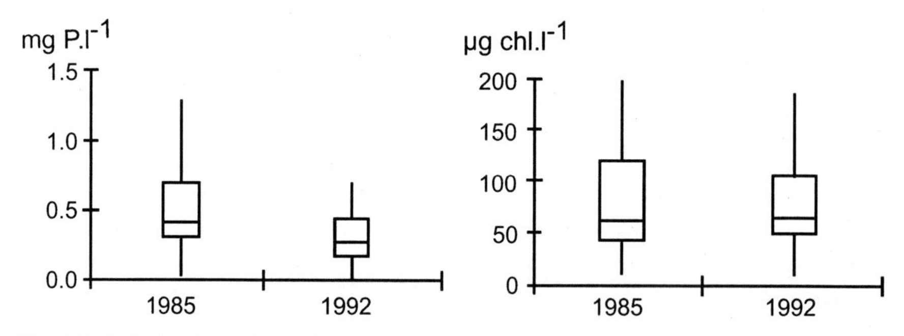

Fig. 1.6. Calculated trends in phosphate levels (left panel) and algal densities (right panel) measured in 1985 and 1992 (RIVM 1995)

Top-Down Control

Daphnids are key organisms for maintaining the clear water state. As long as they are able to consume the algal production facilitated by the increased nutrient concentrations, the system will remain clear. However, anything that, affects the performance of daphnids, either directly or indirectly, may induce a shift to the turbid state. The daphnid community s is influenced by the rate of consumption of individuals by planktivorous fish, which in their turn are influenced through predation by piscivorous fish. The effect brought about by fish on phytoplankton dominance via zooplankton, is known as a trophic interaction or a trophic cascade, as the impact cascades down the trophic levels in the food chain (Scheffer 1998; Carpenter 1993). Top-down control is frequently confirmed in lakes by responses to biomanipulations and changes in plankton communities after fish immigrations or a fish dieoff (Vanni et al. 1990;Vanni and Layne 1997). On the other hand, investigations comparing lakes with different trophic conditions show a positive correlation of phosphorus not only with the phytoplankton biomass, but also with the biomass of the total zooplankton, the crustacean zooplankton, the fish populations, and with harvested fish. These studies support the bottom-up hypothesis (more available nutrients → more algae → more zooplankton → more planktivorous fish → more piscivorous fish) which is based on the idea that the biomass depends on the fertility of the habitat. Year-to-year comparisons within the same lake usually support the top-down hypothesis, whereas comparisons of different lakes tend to support the bottom-up hypothesis (Lampert and Sommer 1997).

However, it should be realised that top-down and bottom-up forces are two sides of the same coin and that they act complementary to each other. The extent to which each of these processes influences the aquatic food web may vary, depending on environmental conditions as well as human perturbations.

In several studies, food web models have been developed with the aim of acquiring more insight into the underlying mechanisms of eutrophication in lakes. The models have been utilised as a tool for the management of eutrophicated water bodies (Jørgensen 1986). The relations between nutrients and phytoplankton growth are quite complex in shallow eutrophicated water systems. Removing nutrient sources will not always lead to an improvement of the water quality in eutrophicated lakes and additional measures are usually needed. An improvement in the water quality of various shallow lakes was observed when combining nutrient reduction programmes with biomanipulation techniques (Moss et al. 1996a; Lathrop et al. 1996). Daphnid grazing has proven to be a crucial process in the top down control of phytoplankton biomass and the maintenance of the water clarity over a number of years, and has therefore been an important aspect in the biomanipulation of eutrophicated lakes over the last few decades (Meijer, 2000). Consequently, much is known about the biology of *Daphnia* and a large number of different models are available for the purpose of describing daphnid dynamics (Hallam et al. 1990; Scheffer et al. 1993).

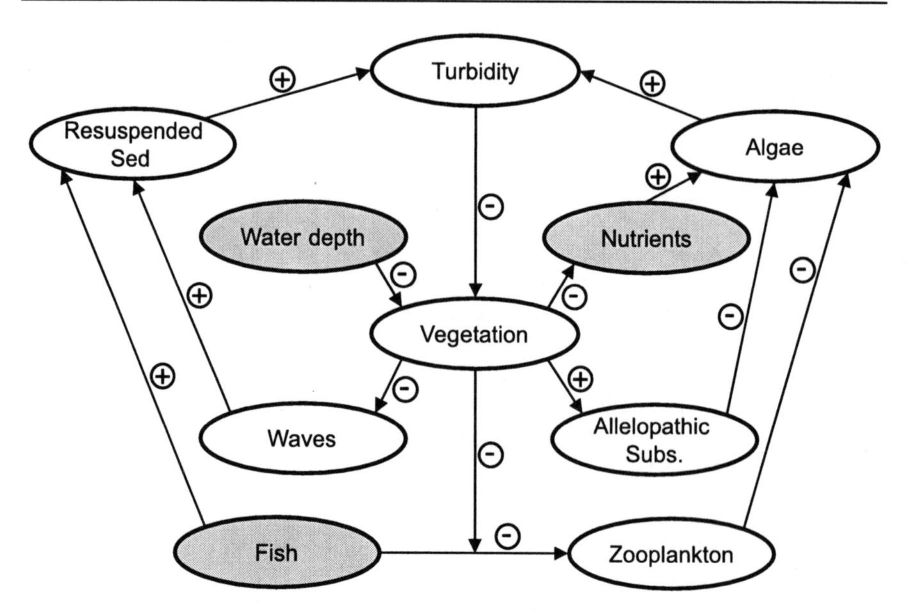

Fig. 1.7. Main feed-back loops thought to be responsible for the existence of alternative equilibria in shallow lake ecosystems. The qualitative effect of each route in the diagram can be determined by multiplying the signs along the way. In this way it can be seen that both the vegetated and the turbid state are self-reinforcing. The shaded boxes are possible steering variables for lake management and the effects of measures can be determined in a similar way (Scheffer et al. 1993; Hosper 1997).

Feedback Loops

The daphnid-algae relationship can be modelled with different types of models. For example, McCauley et al. (1988) and Scheffer (1998) used a simple predator-prey model; its dynamics are affected by factors such as the nutrient level and spatial and temporal heterogeneity. Simulation models for the dynamics of various size classes of daphnids feeding on various size classes of phytoplankton at limiting phosphorus concentrations and the responses of food webs to fish manipulations have also been developed (Carpenter and Kitchell 1993).

Daphnid populations are directly influenced by the availability of food (bottom-up) and by the presence of zooplanktivorous fish (top-down). However, these direct causal relationships are embedded within the freshwater ecosystem that comprises many components which encompass a large number of direct and indirect interrelations. These relations are influenced by biological variables as well as system characteristics, resulting in several positive and negative feedback loops. Scheffer et al. (1993) proposed a schematic representation for such feedbacks (Fig. 1.7). The presence of submerged vegetation is central in this scheme. Vegetation has a direct negative influence on algal growth, by competing for available nutrients and by the release of allelopathic substances. Indirect effects on algal development are offset by the positive effects of vegetation on daphnid populations, since vegetation offers daytime refuge to daphnids and ambush cover for piscivorous fish. By dampening wave action, the vegetation prohibits exces-

rous fish. By dampening wave action, the vegetation prohibits excessive resuspension of sediment which might inhibit its own growth.

Hysteresis

Once a system is turbid, vegetation cannot persist and populations of daphnids and piscivorous fish (e.g. pike and perch) decline. The fish community in turbid waters is comprised mainly of benthivorous (benthos-eating) and planktivorous (zooplankton-eating) fish e.g. bream and roach. The abundant planktivorous fish control the zooplankton, resulting in low grazing of algae and increased recycling of phosphorus. Large numbers of benthivorous fish, if unhindered by vegetation, stir up the mud and thereby contribute to turbidity and phosphorus release (Hosper 1997; Scheffer 1998) and hamper the settlement of new submerged vegetation. These conditions are unfavourable to daphnids, resulting in dominance by less efficiently grazing copepods and rotifers in the zooplankton and prolific algal growth, often dominated by cyanobacteria which are considered to be an inferior food source for daphnids (Moss et al. 1991).

As a consequence, two alternative stable equilibria can occur over a range of nutrient concentrations in shallow lakes: a clear water state, characterised by the dominance of macrophytes, and a turbid water state, in which the phytoplankton is dominant (Scheffer et al. 1993). Below a concentration of ca. 25–50 µg/total P and 250–500 µg/l total N, a shallow lake is likely to be in a clear water state (Moss et al. 1996a). Above this concentration, the water will remain clear, unless the system is disturbed (Scheffer et al. 1993), which might initiate a rapid switch to the turbid state (Fig. 1.8). There is probably no upper limit above which a turbid state only can exist but, at higher nutrient concentrations, the clear water state becomes very unstable (Moss et al. 1996a).

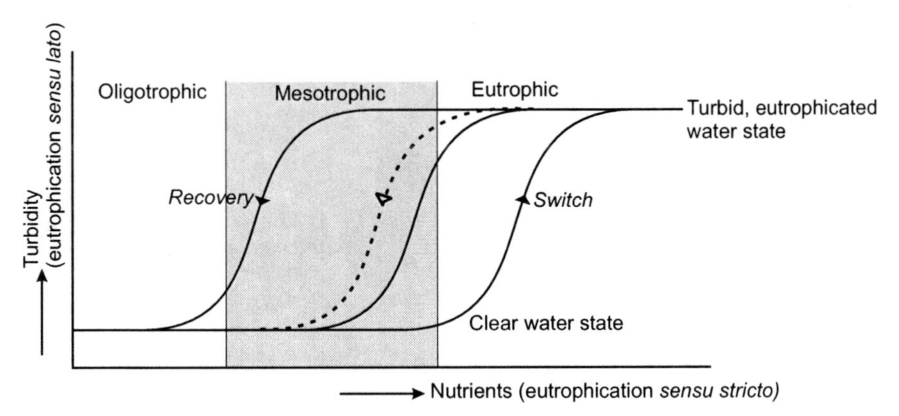

Fig. 1.8. Hysteresis effect in the response of lake turbidity to changes in the nutrient status (adapted from Scheffer et al. 1993)

The presence of two alternative stable states, and the hysteresis that is observed when forcing the system from one stable state into the other, has important management implications, because the way in which the system responds to measures is very different from that of non-hysteretic systems (Hosper 1997; Scheffer 1998).

The turbid water state can be referred to as "eutrophicated water". Figure 1.8 clearly shows that eutrophication problems can occur over a wide range of nutrient conditions suitable for non-eutrophicated clear waters. The model also indicates that restoration from a turbid, eutrophicated state back to a clear water state is not an easy reversion.

Switches

Eutrophicated, turbid, shallow lakes can be resistant to nutrient reduction. The turbid and the clear water state are both hysteretic stable states, as changes in the nutrient concentration within a certain range will not affect the state these lakes are in. Only at certain threshold levels for nutrients do the systems show drastic changes ('shifts') in their algal biomass. Special events or actions ('switches') may trigger a shift from the one state to the other. Buffering mechanisms can prevent or promote switches between the two states for a broad range of nutrient concentrations (Table 1.1 and Table 1.2).

Table 1.1. Stable states, buffering mechanisms maintaining the stable states and switches to trigger a shift from a stable turbid state to a stable clear water state (forward switch) (adapted from Hosper 1997)

Factors contributing to stability of turbid water	Buffering mechanisms	Forward switches to 'clear water'
Oscillatoria bloom	– Resistant to low TP, low light and low temperature – Reduced edibility for Daphnia grazers – Bloom results in high pH, high sediment oxygen demand and thus high internal P loading, more blooms	– Prolonged snow-covered ice – Washout by winter flushing – Control P release by sediment removal, sediment treatment or 'hard water' flushing
Phytoplankton bloom	– Bloom results in turbid waters, low piscivores, high planktivores, low grazing, more blooms	– Lower water level in spring to promote submerged vegetation – Natural winter fish kills – Reduce planktivores and promote piscivores
Non-algal turbidity	– Wind-induced resuspension of sediments in plant-free lakes – Fish-induced resuspension of sediments by benthivores, unhindered by plants	– Reduce wind exposure of sediments or complete drawdown and drying of sediments – Reduce benthivores

Table 1.2. Stable states, buffering mechanisms maintaining the stable states and switches to trigger a shift from a stable clear state to a stable turbid water state (reverse switch) (adapted from Hosper 1997)

Factors contributing to stability of clear water	Buffering mechanisms	Reverse switches to 'turbid water'
Benthic diatoms	– Reduce susceptibility of lake sediments to wind-induced resuspension – Compete with phytoplankton for N,P – Promote N loss by denitrification	– Benthivore stocking – Storm events
Submerged vegetation	– Competes with phytoplankton for N,P – Promotes, N loss by denitrification – Reduces susceptibility of lake sediments to wind-induced resuspension – Excretes substances allelopathic to phytoplankton – Promotes grazing of phytoplankton by providing refuge to Daphnia – Promotes phytoplankton grazing by providing refuge to pike and subsequent top-down control of planktivores – Reduces fish induced resuspension by hindering bottom feeding	– Mechanical destruction of vegetation – Chemicals toxic to vegetation – Macrophyte grazing by birds – Increase water level during spring – Benthivore stocking – Grass carp stocking – Chemicals toxic to Daphnia – Storm events

Disturbance of the Trophic Cascade Due to Toxic Stress

In many shallow lakes in countries such as England, the Netherlands and Denmark, the shift from a clear water state towards a turbid state is thought to have taken place during the nineteen sixties and seventies. The challenge presented in the restoration of a lake is to reverse this shift, and recreate stable clear-water in that lake.

It is worthwhile considering the toxic state of eutrophicated waters, especially where biomanipulation has failed. Toxic stress can trigger a forward shift form a clean water state into a eutrophicated state, as suggested by Hurlbert (1975) and Stansfield et al. (1989), at lower nutrient (trophic) conditions, and may, on the contrary, hamper the recovery of eutrophicated waters back into clear water in response to nutrient reduction (see Fig. 1.9).

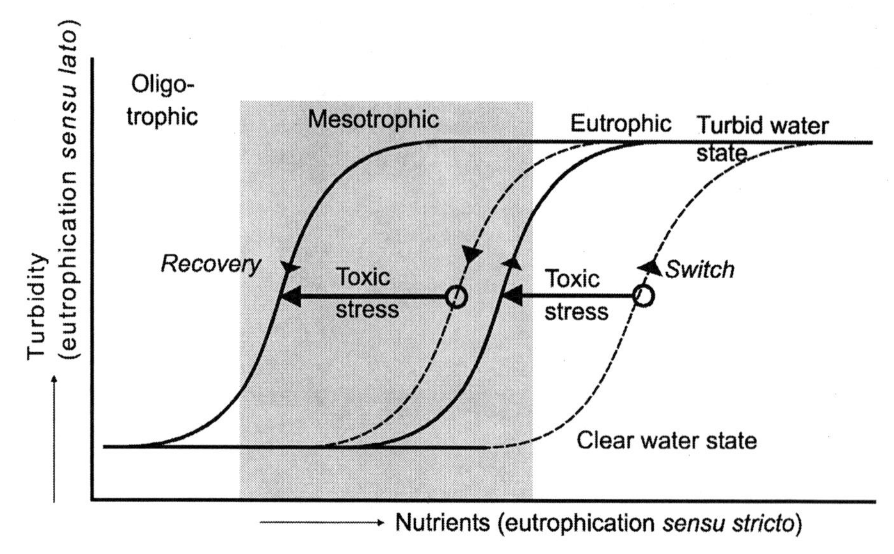

Fig. 1.9. Hysteresis effect in response of lake turbidity to changes in nutrient status (adopted from Scheffer et al. 1993), extended for the impact of toxic stress

Some evidence for reduced daphnid grazing effectiveness due to the presence of pesticides in the field situation is known from literature (see also Chap. 5). Poisoning of the cladoceran (viz. Daphnia) community might have caused a switch from dominance by submerged plants to phytoplankton dominance in the Norfolk Broads, a series of lakes, during the 1950s and 1960s. This allowed phytoplankton to take advantage of the increased nutrient loadings and to increase in number. The 1950s and 1960s were periods of liberal organochlorine pesticide use (Stansfield et al. 1989).

In sediment cores with residues of dieldrin and DDD, no remains of *Daphnia* were found. In addition, the phytoplankton related *Bosmina* rather than submerged macrophyte associated *Chydorus* were found in these cores.

The experimental studies reviewed in this book provide some more insight into and information on the hypothesis that toxic stress may contribute to the shifting of water bodies into eutrophied states, due to disturbances to the trophic cascade at the level of daphnid function.

2 Daphnid Grazing Ecology

2.1 Daphnids and Their Ecological Role

Daphnids play a central role in the food webs of aquatic ecosystems. They often fulfil a key role in making primary production (algae) available for higher trophic levels, including vertebrates (fish) and invertebrate species. Since daphnids and other species within zooplankton communities have high reproduction rates, they are capable of responding to varying levels of algal production. This rapid response enables them to substantially affect the density of algae, which results in low algal densities during part of the season (e.g. the 'clear water phase' often observed in lakes during early summer).

Daphnids Within the Zooplankton

Daphnids form part of the zooplankton. By definition, zooplankton communities are built up of animal species (zoo = animal) that float in the water column (planktos = floating). These species belong to several taxonomic groups (Lehman 1991), but in fresh waters the zooplankton is often dominated by protozoa, rotifers and crustaceans. Protozoa are small animals that feed on bacteria. They usually spend only part of their lives in the water column. The Rotifera (rota = wheel; ferre = to carry) form a separate phylum of generally small, non-segmented animals that filter water with one or two wheel-like ciliary organs (corona). Crustaceans belong to the phylum Arthropoda and have two major representatives in the zooplankton; the class Copepoda, and the sub-order Cladocera of the class Branchiopoda. The ecological differences between the major groups that feed on algae are briefly characterised by Allan (1976, Table 2.1). Some common zooplankton species in European temperate water bodies are presented in Fig. 2.1.

Compared with copepods and rotifers, cladocerans exhibit strong grazing impacts which is related to relatively high growth rates in combination with their relatively large size. A more recent compilation of literature data shows that the variability of the intrinsic rate of increase (r_{max}) within these three groups is much higher than predicted by Allan (Andersen 1997). Rotifers tend to have the highest r_{max}, and copepods the lowest.

Table 2.1. Summary of key ecological characteristics of the major zooplankton taxa (from Allan 1976)

Feature	Rotifera	Cladocera	Copepoda
r_{max} (day^{-1})	0.2 – 1.5	0.2 – 0.6	0.1 – 0.4
Typical adult body size (mm)	0.2 – 0.6	0.3 – 3.0	0.5 – 5.0
Largest species (mm)	1.5	5.0	14.0
Food size range (µm)	1 – 20	1 – 50	5 – 100
Mode of feeding	Suspension feeding via coronal cilia	Filter feeding via thoracic appendages	Filter and/or raptorial
Filtering rate	Very low	High	Low
Susceptibility to vertebrate predators	Very low	High	Low
Susceptibility to invertebrate predators	High	Moderate	Moderate to high
Abundance pattern	Vernal peak	Vernal peak	Variable
Biogeography (large lake distribution)	Surface and inshore	Surface and inshore	Deep, open water
Biogeography (global)	Freshwater and estuarine	Freshwater and estuarine	Freshwater, estuarine, and marine

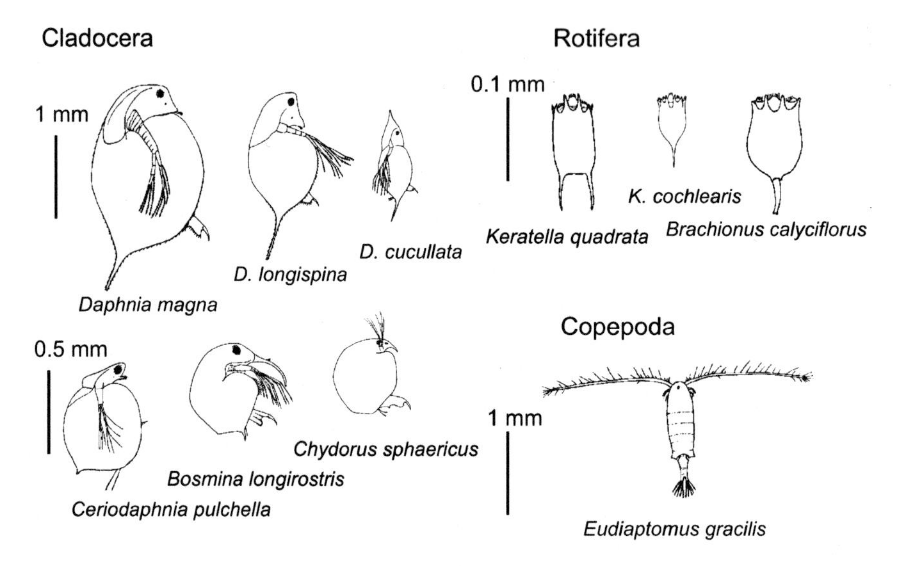

Fig. 2.1. Common zooplankton species in European temperate water bodies (redrawn from Streble and Krauter 1988)

Table 2.2. Overview of cladoceran taxonomy, with special attention paid to the Daphniidae family, and species of the genus Daphnia

- **Phylum Crustacea**
- Class Branchiopoda
  - **Order Cladocera**
    - Suborder Anomopoda
      - **Family Daphniidae**
        - Genera: *Daphnia, Ceriodaphnia, Daphniopsis, Simocephalus, Scapholeberis*
          - Subgenera Daphnia
            - species incl. *D. ambigua, D.obtusa, D. pulex*
          - Subgenera Hyalodaphnia
            - species incl. *D. catawba, D. longiremis, D. longispina*
          - Subgenera Ctenodaphnia
            - species incl. *D. lumholtzi, D. magna, D. similis*
      - **Family Moinidae**
        - Genera: *Moina, Moinodaphnia*
      - **Family Bosminidae**
        - Genera *Bosmina, Bosminopsis*
      - **Family Macrothricidae**
        - Genera incl. *Acantholeberis, Ilyocryptus, Macrothrix*
      - **Family Chydoridae**
        - Genera incl. *Alona, Chydorus, Eurycercus, Pleuroxus*
    - Suborder Ctenopoda
      - **Family Sididae**
        - Genera incl. *Diaphanosoma, Sida*
      - **Family Holopedidae**
        - Genus *Holopedium*
    - Suborder Onychopoda
      - **Family Polyphemidae**
        - Genus *Polyphemus*
      - **Family Podonidae**
        - Genera incl. *Evadne, Pleopsis, Podon*
    - Suborder Haplopoda
      - **Family Leptodoridae**
        - Genus *Leptodora*

Taxonomy

Daphnids comprise a family of the order Cladocera. It has recently been suggested that the order Cladocera is a monophyletic group, divided into 4 suborders, 11 families, about 80 genera, and roughly 400 species (Colbourne and Hebert 1996). The family Daphniidae includes the following genera: *Ceriodaphnia, Daphnia, Daphniopsis, Megafenestra, Simocephalus* and *Scapholeberis*. Many species are planktonic and occur in a wide variety of aquatic ecosystems, such as lakes, canals, and ponds. Other genera with important planktonic representatives are *Moina, Bosmina, Chydorus, Diaphanosoma*, and *Leptodora*, the last being a predator of other zooplankton.

The genus Daphnia is further divided into 3 subgenera, namely *Daphnia*, *Hyalodaphnia*, and *Ctenodaphnia*. Hybrids of species that cannot reproduce sexually, but that are able to reproduce asexually, can be formed. An overview of cladoceran taxonomy and the position of daphnids is presented in Table 2.2. The term 'daphnids' will be used here for species that belong to the family of Daphniidae, and thus includes species from the genera *Daphnia, Ceriodaphnia*, and *Simocephalus*, and can also be applied for species of *Daphniopsis*, *Megafenestra*, and *Scapholeberis*. When other species from other genera (families) are involved (e.g. *Bosmina*, and *Chydorus*), the term 'cladocerans' will be used.

Habitats

Cladocerans typically occur in a variety of habitations, including the water columns of small pools and very large lakes; the littoral zone, where many species attach themselves to aquatic plants and reed stems, and the upper layer of sediments. Daphnids can be found in various water bodies ranging from tidal ponds to large lakes. They are often found abundantly in standing waters throughout the world. Although many species are present on different continents, some species are confined to small areas. Daphnids generally feed upon algae, detritus and microorganisms (see Sect. 2.2), which are either filtered from the water or grasped from substrates (e.g., sediment grains and plants). Since daphnids have a high filtering capacity and grow and reproduce quickly, they can respond more rapidly to increases in food availability than filter feeders that live for several years (e.g., bivalves), which can only respond by increasing their individual body sizes. Some daphnid species (e.g., *Daphnia magna*) are very resistant to variations in environmental conditions, such as water temperature, low oxygen concentrations and salinity. Furthermore, the production of winter resistant eggs enables them to recolonise after periods of extreme conditions that cannot be survived by either adults or juveniles.

Predation

Daphnids are vulnerable to predation by fish, due to their relatively large size and slow swimming behaviour, and serve as an important food source for planktivorous fish. In order to avoid predation, species that occur in the pelagic zone tend to be transparent, thereby reducing their visibility, or form morphologic adaptations, such as helmets (known as cyclomorphosis), so as to reduce their susceptibility. Another adaptation that is likely to be induced by predation is the diurnal vertical migration observed in many lakes. In fact, they reduce their visibility by migrating to deeper water layers during daytime, and only migrate to the upper water layers, which are rich in algal food, at night.

Reproduction

Reproduction in cladocerans can be both sexual and asexual. During summer, when conditions for growth are optimal, rapid population increase is possible through parthenogenesis, which enables females to produce broods of new females that also reproduce asexually. Eggs are formed in a brood pouch on the ventral side of the carapace. The number and sizes of the eggs vary between species (Lynch 1980), with food availability, temperature and the size of the females taken into consideration. Size increase in individuals is restricted to the period just after moulting when the old carapace is thrown off and water is taken up in order to stretch the body and the new carapace. When environmental conditions become less favourable, due to lack of food, decrease in temperature or oxygen concentration, overcrowding or accumulation of metabolic wastes, eggs are produced from which sexually vital males and females develop. When eggs are fertilised and released into the brood pouch, an ephippium is formed containing dormant embryos (see Zaffagnini 1987). These dormant eggs can resist detrimental conditions, like freezing and drought, and facilitate species survival during unfavourable periods. Since ephippial eggs float on the water surface they are often blown inshore, facilitating dispersal. Only parthenogenetic females hatch from ephippial eggs, which form the inoculum for the new population in spring.

Filtration

Daphnids and other cladocerans combine a relatively high filtration rate with a low critical food concentration threshold (a few µg chlorophyll-a per litre), compared to the other dominant taxa in zooplankton communities. This results in efficient algal grazing and the capacity to control algae at low densities. Moreover, they can easily respond opportunistically to changes in algal production by reproduction under conditions of high food supply. The population density of cladocerans is, therefore, a good reflection of the preceding algal production.

When present in aquatic ecosystems, daphnids have a significant impact on the abundance of phytoplankton. In the northern temperate regions, daphnids graze heavily upon the phytoplankton spring bloom and are responsible for the so called clear-water phase. Food levels are low during this phase and the abundance of zooplankton decreases due to lack of food and/or predation by fish.

Cladocerans (with the exception of *Bosmina*) when offered a mixture of synthetic particles and algae of the same size will ingest the algae and the artificial particles at the same rate. Unlike filter feeding cladocerans and rotifers that select their food primarily on the basis of size, raptorial herbivores are capable of selecting food based on its chemical qualities ("taste"). Herbivorous copepods can 'test' each particle and decide whether they will eat it or not (Lampert and Sommer 1997).

While having a relatively low threshold food level, indicating the minimum food level required to maintain a positive population growth, cladoceran species reach relatively high population growth rates (see Fig. 2.2), compared to other dominant taxa in zooplankton communities (copepods and rotifers). Although the variability between species within the different groups is high, species with high threshold food levels tend to have high maximum growth rates. Threshold food levels can be considered as a measure of the competitive ability of species during low food conditions. Therefore, copepods will, tend to become dominant when food levels are low. In the case of high algal production rates, however, the low intrinsic rate of copepod increase, combined with a relatively low biomass, may become too low to keep up with the algal production, thereby resulting in high

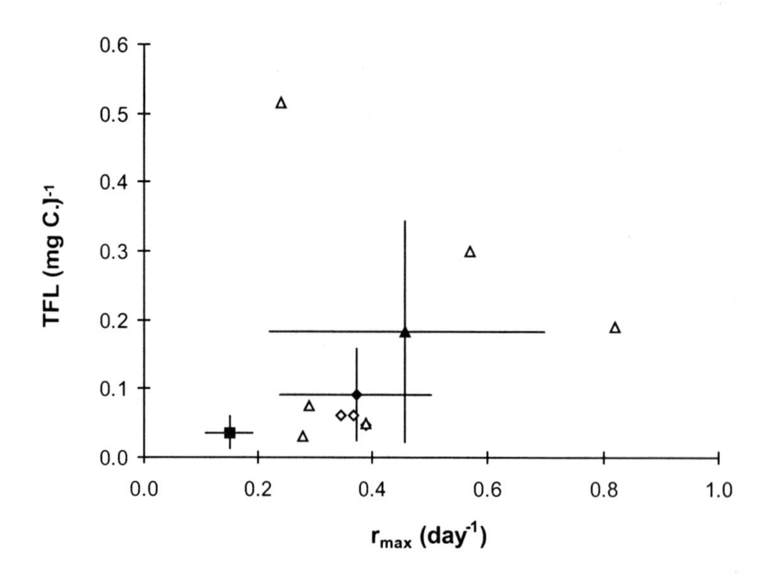

Fig. 2.2. Threshold food levels (TLF) for nett population growth related to maximum rates of increase (rmax) for cladocerans (8 species; ◊), copepods (5 species; □), and rotifers (11 species; Δ). Open symbols refer to individual species, filled symbols and lines indicate averages and their standard deviations. Data from Andersen (1997)

algal densities. Cladocerans, having higher rates of increase and a higher individual biomass, are more effective in their response to varying food levels.

Feeding Rate

In common with most cladocerans, daphnids are mainly filter feeders that collect particles from the water by using their thoracic appendages. The volume of water from which particles are removed, per unit of time, is referred to as the filtering rate, filtration rate, clearance rate, or grazing rate (Lampert 1987a). The food that is collected is passed from appendage to appendage and moved along a ventral food groove propelled by the same thoracopod movements that filter the food from the water. Finally, the food reaches the mouth that is situated at the underside of the head close to the carapace, the latter enclosing the thorax and abdomen. The amount of food that passes the mouth and enters the gut over a certain period of time is referred to as the ingestion rate or feeding rate (Lampert 1987a). The C-shaped intestine passes through the thorax and abdomen, and ends at the distal end of the post-abdomen where the anus is situated.

Basically, all particles of suitable size are ingested. For daphnids, the appropriate particle size ranges from 1 µm (bacteria) in diameter to particles with a diameter of 70 µm for large (3mm) *Daphnia magna* (Burns 1968). The upper size range (Y) increases with body size (X), following the relationship: $Y=22X+4.87$ (Burns 1968). However, natural food particles are rarely spherical, and the shapes of particles also determine their suitability as a food source. Particles are ingested without selection, and the sizes gathered are set by the filtering mesh constructed by the thoracic appendages. Material that is too large or undesired can be removed from the mouth by the first legs. Suitable food sources include algae, bacteria, and protozoans.

The factors exerting a major effect on the filtering and ingestion rate are the concentration of food, food type, body size and temperature. Other factors, such as the degree of hunger, time of the day, light conditions and stress factors can also have an effect on these rates (Lampert 1987a). The ingestion rate increases with the concentration of food when food levels are below a level called the incipient limiting level (ILL; McMahon and Rigler 1965). Above this level, the ingestion rate does not increase further, the filtration rate declines and the excess of collected food is rejected. The highest specific ingestion rates are in the order of 3% of the animals' body carbon per hour. The filtering or ingestion rate R increases with body length L, following $R=a.L^b$, where a and b are constants with b often between 2 and 3. Knoechel and Holtby (1986) presented a regression equation from which the community filtering rate can be derived from the rate (F, $ml.animal^{-1}.day^{-1}$) per individual with a length L in mm: $F=11.695L^{2.48}$.

Thus, with an increase in body length, the ingestion rate increases exponentially. The response of filtering and ingestion rates to temperature is described by an optimal curve, for which the optimum temperature depends on the species. A lowering of the temperature would decrease these rates with Q_{10} values of 1 to 3. The response of filtering, ingestion and rejection rate to different food levels is illustrated in Fig. 2.3.

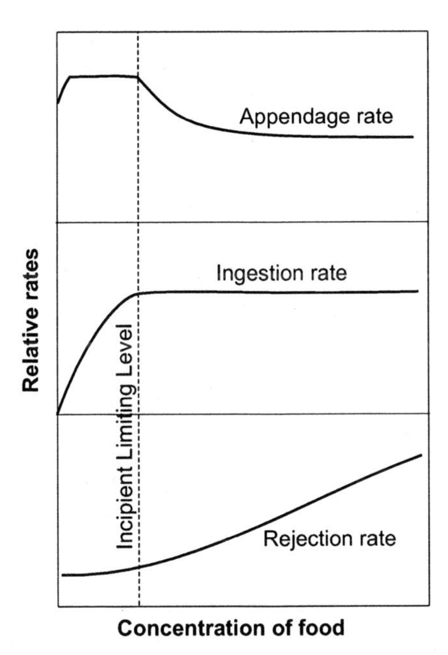

Fig. 2.3. Beat rates of appendages, ingestion rates and rejection rates as a function of food concentration. The broken line indicates the incipient limiting level. Redrawn from Lampert (1987c)

Digestion

Only a fraction of the food is efficiently digested and assimilated. Assimilation efficiency varies considerably between food sources of differing qualities, and reported calculated values range from about 10% to 100% with most values being lower than 60% (Lampert 1987a). The assimilation efficiency increases with temperature and decreases markedly with increasing food concentrations. The assimilated food is used for maintenance, growth, reproduction, development of body features such as helmets, storage and production of substances in response to changes in environmental factors.

Growth and reproduction is mainly regulated by the amount of available food resources and temperature. Different species have different strategies with regard to the allocation of their assimilated food. Some species tend to grow and mature rapidly and produce many small eggs per brood, while others grow slowly, mature at a relatively old age and produce a limited number of usually larger young. Within one species, these characteristics show some variability in response to food concentrations, with the production of fewer eggs that develop into large young when food levels are relatively low (Enserink 1995). Since different life stages may respond differently to food levels, a critical stage in development may occur that determines the minimum required food level (Goulden et al. 1982). This critical life stage varies between species (Matveev and Gabriel 1994), but the juvenile stage is usually the most sensitive to starvation.

Food Selection

Cladocerans are known as non-selective filter feeders. However, not every particle in the seston is a suitable food source. The feeding efficiency of cladocerans may be inhibited by particles of unsuitable size and shape (i.e., filamentous structures and other colonial algal forms), whereas (bio)chemical composition may have a pronounced influence on the assimilation efficiency of the ingested food. Algae, which are often the most important food source for cladocerans, have the ability to reduce their palatability in response to grazing pressure. Flagellates and many diatoms are generally considered to be a qualitatively good food source for daphnids. Cyanobacteria are considered an inferior food source, while green algae take an intermediate position (Gulati and DeMott 1997; Lürling and Van Donk 1997).

Single cells with a maximum diameter of ca. 35–45μm are probably the most optimal food items in the matters of size and shape (Lampert 1988; Zurek and Bucka 1994). Larger cells will be rejected, but elongated cells may be eaten. Colonies of cells can be broken down and the individual cells ingested. However, this will increase the time required for feeding and consequently reduces nett feeding efficiency. Interference with feeding efficiency (e.g., formation of large colonies), is in fact one of the strategies used by edible algae in response to high grazing pressure (Lampert and Sommer 1997).

The most important biochemical constraints determining the quality of algae as a food are their nutritional composition. Amino acids and fatty acids are important dietary components which can only partially be synthesised by the animals themselves. Cladocerans probably need to extract as many as 10 amino acids from their food (Harrison 1990). Recent developments in the elemental stoichiometry of freshwater ecosystems indicate that the ratio of the macroelements carbon, nitrogen and phosphorous in the seston is a very good indicator for the quality of the food (Elser and Urabe 1999). It is generally thought that algae containing these nutrients in the molar ratio 106C:16N:1P (Redfield ratio) grow under nutrient sufficient conditions (Hillebrand and Sommer 1999). Rapidly growing algae are a qualitatively good food source for cladocerans, especially for daphnids, which have a very similar nutrient ratio (Sommer 1992). The nutritive quality of algae will decrease, due to changes in the nutrient ratio of the algae under nutrient limited conditions. On account of their specific nutrient stoichiometry, copepods are specifically favoured in systems dominated by slightly P-deficient algae (Sommer 1992). However, many algae are able, to a certain extent, to concentrate nutrients (especially P) in response to temporary shortages (Hessen 1990; Sommer 1989). Under P limited conditions, algae may become an even more favourable food source for daphnids. However, many algae respond to nutrient deficiencies by increasing their grazing resistance (Butler et al. 1989; Sterner 1993; Van Donk and Hessen 1993; Lürling and Van Donk 1997; Van Donk et al. 1997).

The presence of toxins in cyanobacteria is a direct biochemical constraint to their quality as food (Haney et al. 1994). However, direct toxic effects from cyanobacteria are only rarely observed and large densities of daphnids may be observed during cyanobacteria blooms (De Bernardi and Guissani 1990). Most frequently, the grazing resistance of cyanobacteria is likely to be the result of morphological constraints (e.g., the formation of large aggregates, which reduces feeding efficiency). Their high P storage capacity may even render them a necessary supplementary food source for cladocerans when more edible algae are P-limited (DeMott 1998).

Life History Strategies

According to Romanovsky (1985), three extremes in life history strategies to do with cladocerans can be distinguished on the basis of their body size, reproduction effort and the resistance of juveniles to food deficiency (Fig. 2.4). These characteristics may help to explain the success of different species under different conditions of food availability. The vulnerability of juveniles to low food conditions is a crucial point in their life history strategies. In addition to food availability, predation has a striking effect on the life history characteristics of cladocerans (Lynch 1980).

Species in Low Productivity Ecosystems

Dominant cladoceran species in systems with low trophic status and small seasonal variations are small and slow growing, so that only minor growth is required before reaching maturity. They produce a small number of relatively large eggs, which enhances the survival of the juveniles at low food levels. Small species are not vulnerable to fish predation, but may be consumed by invertebrates (e.g. insect larvae). Some typical examples are Bosmina coregoni, and Diaphanosoma brachyurum.

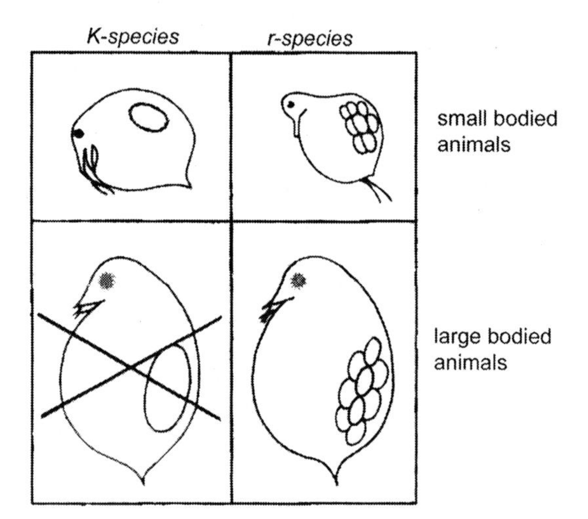

Fig. 2.4. Life history strategies in cladoceran crustaceans (modified from Romanovsky 1984)

Species in Eutrophic Ecosystems with Varying Food Levels
Species with high birth rates will benefit most in conditions of high food availability. Many small eggs are produced, requiring that juveniles grow substantially before reaching maturity. This makes the juvenile stage vulnerable to low food levels. When food becomes depleted, however, large-sized adults are able to survive for a relatively long period of time, since the metabolic losses per unit of mass are lower for large bodied individuals. These large adult individuals are, however, very susceptible to fish predation. Some typical examples are Daphnia magna, Daphnia pulex, Daphnia galeata, and Simocephalus vetulus.

Species in Eutrophic Ecosystems
When high food levels persist (e.g., because large daphnids become excluded by fish predation), smaller species with high birth rates may become dominant. They have an intermediate egg size and require only a short development time. In order to optimise egg reproduction rates, females continue to grow after reaching maturity, thereby increasing the number of eggs produced per brood. Their short lifespans, and small size, do not enable them to survive periods of low food availability. In this situation, smaller species are at a competitive disadvantage because of their higher respiration to assimilation ratio (Gliwicz, 1990). Some typical examples are Ceriodaphnia reticulata, and Daphnia cucullata.

Community Structure

The composition of the zooplankton community is important for the qualitative and quantitative impact on algal assemblages. Large species can play a role in responding to locally or temporarilly high algal productivity, while small, slow growing species can benefit from the competitive advantage of long-lasting low food levels. Since smaller zooplankton species cannot consume larger algal species, these larger algal species might be at an advantage when grazing rates are high. Small rapidly reproducing species fill the niche in between. A natural assemblage of various daphnid species, with populations including various life stages, may thus have a higher grazing efficiency than a monospecies/single cohort population.

Since the grazing rate of individual zooplankton is related to its size, the filtering rate exhibited by the total (cladoceran) community can be estimated from its size structure (Knoechel and Holtby 1986). Although the filtration rate is influenced by factors such as the time of day, type of food, concentration of food, oxygen concentration and the presence of filamentous cyanobacteria, the calculation of the community grazing rate gives an indication of the overall grazing pressure on the phytoplankton community (Thompson et al. 1982).

Seasonal Succession

The seasonal succession in plankton dynamics is influenced by temperature, light and the availability and concentration of nutrients. The international Plankton Ecology Group (PEG) developed a model to illustrate the course of seasonal succession in 24 sequential events (Fig. 2.5). Seasonal succession in eutrophic lakes has two phytoplankton maxima, i.e.a spring maximum of small algae, and a summer maximum of large, grazing resistant forms.

These maxima are separated by a clear-water phase. The clear water phase is controlled by a maximum of large zooplankton that is later replaced by smaller zooplankton species. The seasonal pattern is different in oligotrophic lakes. (Sommer et al. 1986, Scheffer; 1998).

Field observations of 13 eutrophic Wisconsin (U.S.A.) lakes dominated by either the larger bodied *Daphnia pulicaria* or the smaller bodied *Daphnia galeata mendotae* showed a higher biomass and filtration potential when *D. pulicaria* was the most dominant species in the zooplankton. *D. pulicaria* appeared to delay algal bloom conditions, but it did not ultimately prevent the algae from growing. Nevertheless, the clear-water phase started earlier, lasted longer, and was usually characterised by greater Secchi disc readings than was the case for *D. galeata* dominated lakes. High densities of large-sized Daphnia could be preferable for biomanipulation because they can attain filtration potentials high enough to increase summer water clarity in eutrophic lakes (Kasprzak et al. 1999).

Impact of Daphnid Grazing on Algal Dynamics

In temperate latitudes, clear seasonal patterns are observed in plankton dynamics. These patterns are related to variations in solar radiation and temperature, which determine the light availability for primary producers and the metabolic activity of zooplankton and other animal life. A simplified model of the major seasonal planktonic events in freshwater lakes in the temperate zone is described by Sommer et al. (1986). In general, phytoplankton density increases during spring as a result of increased solar radiation. This spring peak is heavily grazed upon by zooplankton, mainly cladocerans, which respond in an opportunistic manner to the enhanced availability of food. As a result, the spring peak is followed by a clear-

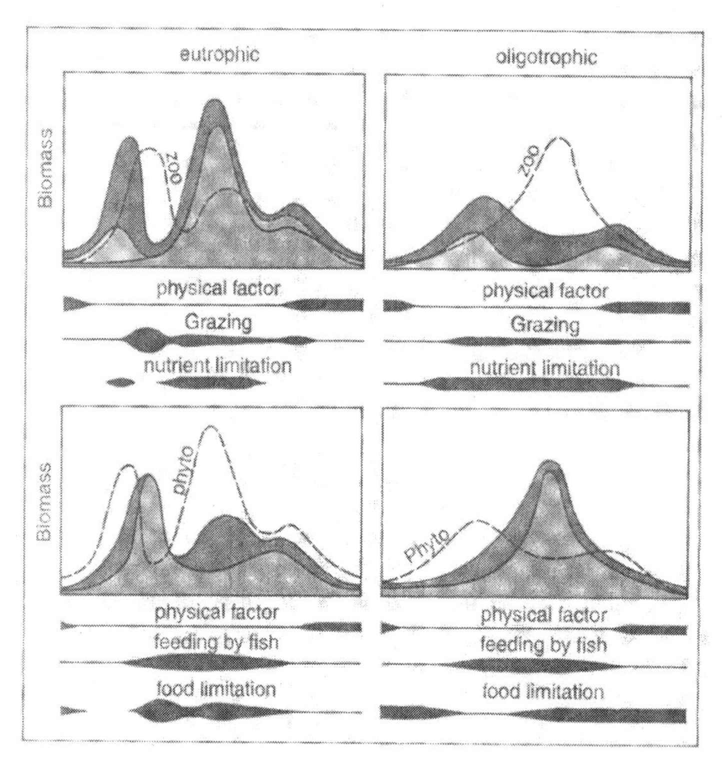

Fig. 2.5. Graphic presentation of the PEG model of seasonal succession. Seasonal development of phytoplankton (above) and zooplankton (below) in eutrophic (left) and oligotrophic (right) lakes. Phytoplankton: dark shading, small species, medium shading, large non-siliceous species; light shading, large diatoms. Zooplankton: dark shading, small species; medium shading, large species. The black horizontal symbols indicate the relative importance of the selection factors (from Sommer et al. 1986)

water phase (Lampert et al. 1986). Later, a decline in zooplankton density is caused by both food limitation and an increase in predation pressure by small fish. This general pattern may, however, vary considerably between lakes depending on flushing rates and the presence of thermal stratification, which is absent in shallow lakes (Sommer et al. 1986).

The impact of zooplankton on phytoplankton development and abundance has been studied by several researchers in field experiments in which zooplankton was excluded by either manipulation by 'natural' factors (e.g., removal of zooplankton, introduction of planktivorous fish) or by the application of a pesticide. Sarnelle (1992) compiled data from several enclosure and whole-lake experiments in order to compare algal responses to nutrient enhancement in the presence of various Daphnia densities. This data shows (Fig. 2.6) that the algal level can be controlled by zooplankton, even in situations of high total phosphorous levels. An easing of zooplankton grazing pressure on algae was shown to result in high to very high chlorophyll-a levels. In other words, top-down control of phytoplankton levels occurs even with a high nutrient loading level, unless the *Daphnia* density is low.

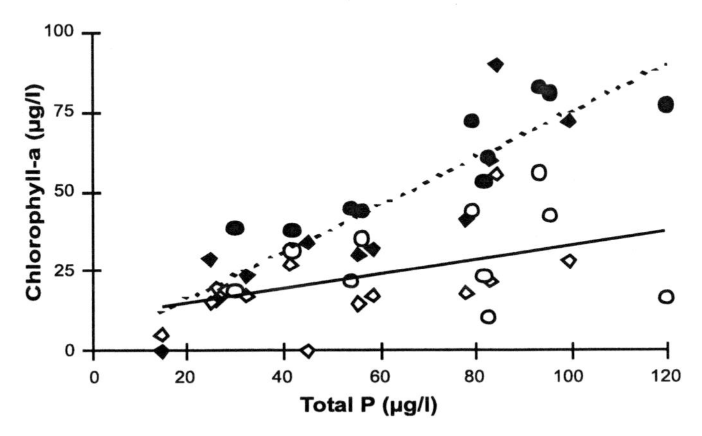

Fig. 2.6. The relationship between the P- and chlorophyll a-levels in lakes with low (solid symbols, broken line) and high (open symbols, solid line) Daphnia densities. Circles indicate that chlorophyll levels were calculated from total cell volume (Redrawn from Sarnelle 1992)

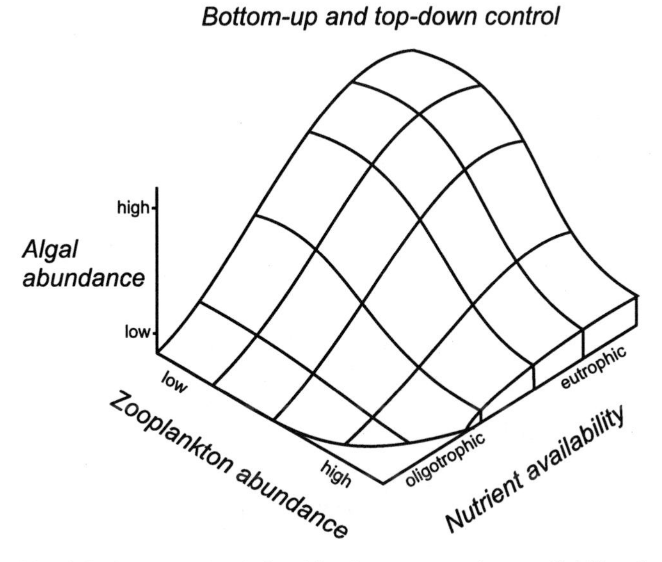

Fig. 2.7. Conceptual relationships between nutrient availability (bottom-up control), zooplankton abundance (top-down control) and phytoplankton abundance in lakes (from Spencer and Ellis 1998)

Based on the results of several studies, Spencer and Ellis (1998) constructed a conceptual relationship between nutrient availability, zooplankton abundance and phytoplankton abundance in lakes (Fig. 2.7). At low nutrient levels, algal concentrations are always low, and zooplankton development is minimal. With increasing nutrient availability, the algal abundance may be high or low, depending on the abundance of zooplankton. This top-down control of phytoplankton abundance via zooplankton grazing is typically mediated by cascading trophic interactions produced by alterations in the fish community and / or other zooplankton predators (Spencer and Ellis 1998). The abundance or the functioning of zooplankton may be controlled by other factors as well, such as the water silt content, acidity, salinity, or the presence of toxic substances.

Mazumder and Lean (1994) manipulated the density of large cladoceran grazers (Daphnia > 1 mm) by either excluding or adding planktivorous fish. Four large enclosures were used, each with a diameter of 8 m and a depth of 11 m. Both treatments were duplicated. Nutrients were added weekly, and the system dynamics were followed for almost four months. Most of the added nutrients accumulated in the epilimnion. In the systems without large daphnids, most of the phosphate was present in the particulate fraction, and small amounts were maintained in the dissolved fraction. Phosphate in particulate matter accumulated almost entirely in algae of an edible size (0.2–20 μm), while the proportion of phosphorous in zooplankton (> 200 μm) formed only a small percentage, even in the absence of fish.

The most striking effect of the nutrient additions was the high rate of phytoplankton development in the systems without large daphnids, and the enduring low densities in the systems were daphnids were present (Fig. 2.8). While nutrient additions resulted in accumulation in the water column, daphnid response was able to control phytoplankton development in the absence of planktivorous fish.

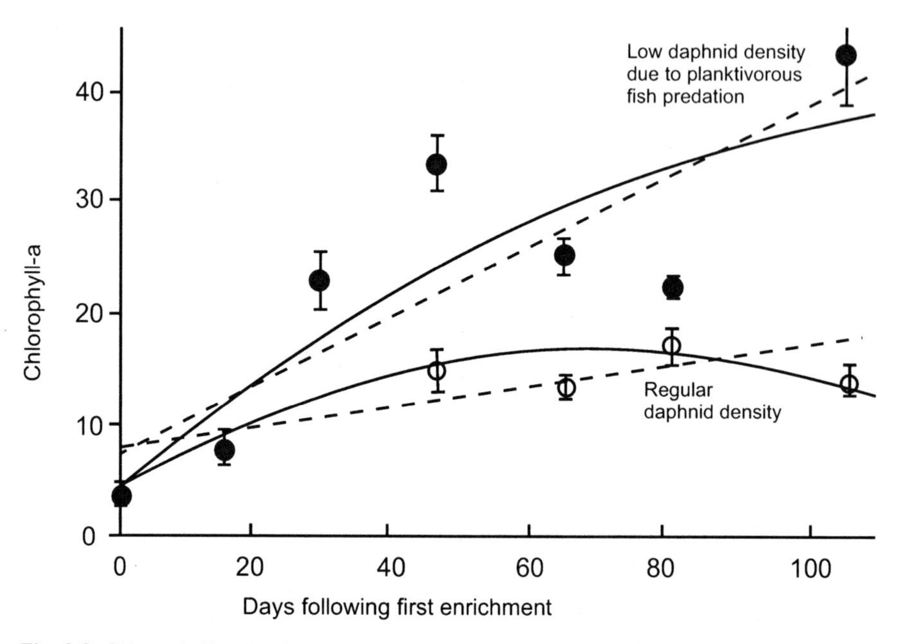

Fig. 2.8. Chlorophyll-a development in enclosures spiked with nutrients under two different conditions with respect to daphnid development; from Mazumder and Lean (1994)

Many studies show that when the zooplankton biomass is high, algal biomass is low (e.g., Mazumder 1993; Mazumder and Lean 1994; Burns and Schallenberg 1998; Spencer and Ellis 1998). Planktivorous fish are often identified as the controlling factor for zooplankton biomass. Since planktivorous fish are predated by piscivorous fish, a cascading effect would affect the structure of the pelagic community. Many studies of this top-down, or predator controlled, organisation have also been carried out (e.g., Carpenter et al. 1987; Mazumder and Lean 1994; Jeppesen et al. 1996; Vanni and Layne 1997). These studies have shown that the presence of piscivorous fish results in lower algal densities at a certain lake trophic level. In a recent study, however, covering 28 Canadian lakes, Currie et al. (1999) were not able to relate either chlorophyll-a levels or zooplankton biomass to the presence of piscivorous fish, although zooplankton biomass clearly affected phytoplankton abundance. Despite this top-down control of the algal biomass, its level was strongly related to total phosphorous, while the zooplankton biomass was not. In lakes with higher total P, the size of the zooplankton, including cladocerans, was smaller. This suggests that size-related predation by planktivorous fish was more significant in these richer lakes than in oligotrophic lakes. Unfortunately, the abundance of planktivorous fish was not studied by Currie et al. (1999), nor was the quantitative importance of predation by fish estimated. The data presented, however, suggest that fish predation in nature might be too low to contribute to cascading effects down to the level of phytoplankton abundance. Future studies should include the quantitative importance of both resource status (total P), and abundance of planktivorous and piscivorous fish in order to clarify their impact on phytoplankton biomass.

Daphnid Grazing in Temperate Lakes

Daphnid grazing in shallow lakes has been studied intensively because shallow lakes are numerous, due to geological processes and excavation by humans. Shallow lakes differ from deep ones by the intense sediment-water interaction and the potentially large impact of aquatic vegetation (Moss et al. 1996a). Depth is the key factor for the processes and characteristics of the different types of lakes. The lake depth influences the sinking loss and resuspension of nutrients but also has very pronounced implications for the light climate experienced by algae, and consequently for their growth rates and realised biomass.

Field data show that enhanced nutrient levels, e.g. phosphorus, can lead to intensive algal blooms in the shallowest lakes, while chlorophyll concentrations soon level off in the deeper ones and become largely independent of the nutrient concentration. Resuspension of sediment particles can contribute to any background turbidity that may have an effect on algal biomass.

The definition of a deep lake is a lake with a depth of more than 10 m that cannot be mixed and overturned throughout. Only deep lakes can be stratified because of depth-related differences in the densities of the layers. Three types of layers can be defined for a stratified lake: the epilimnion, which is mixed throughout by wind and wave circulation; the deeper high density water or hypolimnion, which, except in tropical lakes, is usually much colder; and a fairly sharp gradational zone between the two layers which is defined as the metalimnion.

Besides nutrient availability for phytoplankton (bottom-up control), top-down control of phytoplankton by daphnid grazing differs between a shallow and a deep lake. In shallow lakes dominated by vegetation, grazing may play a major role throughout summer because the plants provide refuge for a very large daphnid population despite an abundance of zooplanktivorous fish. Algal dynamics are less influenced by daphnid grazing in deeper lakes. The chlorophyll-a concentration in stratified West Midland (UK) lakes with a depth > 3m showed a strong correlation with inorganic N concentrations. Grazer control in the shallower West Midland lakes (depth < 3m) was probably linked to the dominance of submerged macrophytes and the refuge they provided for grazers, as daphnid abundance was strongly correlated with chlorophyll-a concentrations. The differences between both types of lakes were probably related to theirmorphology (Moss et al. 1994).

Daphnid Grazing in Subalpine Lakes

Large, subalpine lakes, such as Lake Garda (Italy), are characterised by great depths and large volumes. In the epilimnion of these oligomictic lakes, light, temperature and oxygen are non-limiting factors for primary production. Algal blooms can occur annually in summer when nutrient levels are high. Respiration and decomposition of organic matter are the main processes in the hypolimnion. This part of the lake functions as a sink for nutrients which can be released to the epilimnion when complete internal mixing of the entire lake takes place. This occurs mainly during very cold and windy winters. Grazing seems to be less important in these lakes, except for the epilimnion in more shallow lakes (Salmaso et al. 1999).

Daphnid Grazing in Sub-antarctic Lakes

In sub-Antarctic lakes, the pelagial grazer community consists mainly of copepods. Predation on zooplankton is negligible in these lakes due to the absence of fish. Hence, the phytoplankton in these lakes suffers from intense grazing pressure from large copepods combined with low temperatures and a short growing season.

Field studies in high and low productive South Georgian lakes have shown that the concentration of available phosphorus (PO_4-P) in the water decreased with copepod abundancy, due to inefficient nutrient regeneration via copepods. A likely explanation is that copepods, in contrast to cladocerans, excrete nutrients in well-wrapped fecal pellets that rapidly sink to the sediment surface, making ingested nutrients unavailable to organisms in the water column. Hence, in lakes where copepods are the dominant grazers, algae suffer both directly from grazing and indirectly from reduced nutrient availability (Hansson and Tranvik 1997).

Daphnid Grazing in Tropical Lakes

Some specific features of tropical lakes are: high recycling rates, an elevated primary production throughout the year, a high nutrient assimilation, a high settling velocity for nutrients, an intense organic matter decomposition and high grazing rates. This results in the establishment of a very dynamic ecosystem. Due to these characteristics, warm water lakes react completely differently to the impacts of eutrophication (sensu stricto) than do lentic systems in temperate regions. Moreover, an almost permanent mixing process exists in shallow tropical lakes, which allows for the resuspension of nutrients and a consequent increase in the primary productivity. It can be concluded from studies in Brazilian Lakes and reservoirs that nutrient mass is not often directly linked to the phytoplankton growth because of the high nutrient assimilation in these organisms (Von Sperling 1997).

Zooplankton composition, abundance and distribution have been shown to be indicative of the increasing eutrophication of the Lake Catemaco (Mexico). Indicators of the rapid progress of eutrophication include the low density and diversity, the small size of the zooplankters, the presence of a significant number of indicator species, and the relatively high density of calanoid copepods compared to other planktonic crustaceans (Torres-Orozco and Zanatta 1998). The highly productive tropical systems seem to react more rapidly to eutrophication processes than lakes in more temperate regions. Similarly, reduced zooplankton grazing easily results in enhanced algal densities.

Daphnid Grazing in Rivers

Zooplankton dynamics are regulated by variations in the residence time of the water and the flow velocity. Slow flowing sections of a river with a medium water level are likely to be the most favourable zones for zooplankton. The flow velocity of the water in regulated rivers is often too fast to enable substantial zooplankton development in the main channel (Reckendorfer et al. 1999). Gosselain et al. (1998) suggested that the composition of the phytoplankton community in large rivers may at times be controlled by grazers, as was the case in the Belgian section of the Meuse throughout the growing seasons of 1994–1996. This only takes place when physical constraints are reduced (i.e., when discharge is low) and when high temperature and availability of grazable algae allow high zooplankton biomass.

Potential Grazing Pressure

The influence of zooplankton on eutrophication problems in Dutch lakes was one of the new topics covered in the Netherlands Fourth National Eutrophication Inquiry (Portielje and Van der Molen 1997b). The Potential Grazing Pressure of daphnids was calculated as the ratio between daphnid biomass (mg/l C) and the phytoplankton biomass (mg/l C). This analysis indicates that top-down control of algal biomass is important. The maximum chlorophyll-a: total P ratio (The "Vollenweider" regression) decreases drastically with the increasing Potential Grazing Pressure of the daphnid community.

2.2 Daphnid Grazing Effectiveness – Concept and Measurement

Daphnid grazing effectiveness, in terms of an accurate, timely and synchronised response to (changes) in algal production, is a key factor in the functioning of aquatic ecosystems. The presence of daphnids in relation to the algal density, referred to as "Potential Grazing Pressure" (Matveev and Matveeva 1997; Portielje and Van de Molen 1997b), is no more than an indication of what can be obtained from recording in the field. Whether or not the potential grazing pressure results in actual and adequate grazing also depends on the ambient conditions and the state of the daphnid community that is present. The grazing rate and efficiency of daphnids depends on species, density, age and physiological condition and on algal composition.

The response of daphnids is especially critical during the period during which the algal community is able to increase its growth rate on the basis of the available nutrients (no nutrient depletion) during improving algal growing conditions (light, temperature etc.), such as in the case of seasonal algal bloom in temperate regions (Sommer et al. 1986).

The algal biomass tends to develop exponentially during such a period, and a fast and an efficient grazing response by daphnids is necessary in order to slow down this exponential algal growth. In mesotrophic and eutrophic waters, effective daphnid control can mean the difference between the transition of an algal bloom phase to a clear- water phase or an extended algal bloom (see Fig. 2.9), with a consequent risk of a switch in the water system towards a more persistent turbid, eutrophicated state (according to Moss et al. 1996a and Scheffer 1998).

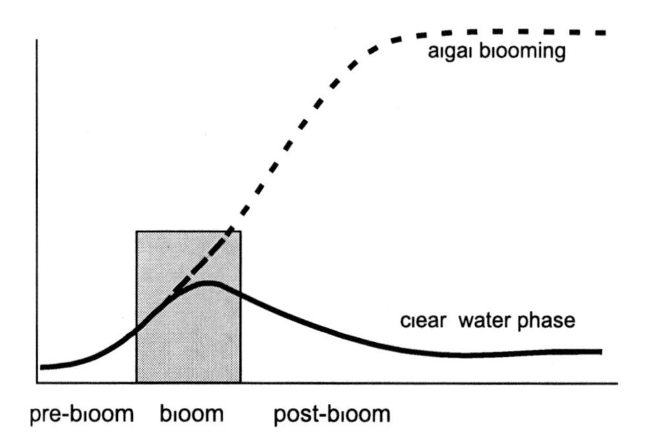

Fig. 2.9. Critical period for the control of algal development by daphnid grazing. Effective daphnid grazing response will control and reduce the algal density down to a clear water phase, while ineffective daphnid grazing can result in extended exponential growth of algae towards a turbid water phase with a risk for the water system of a switch to a more persistent eutrophicated state

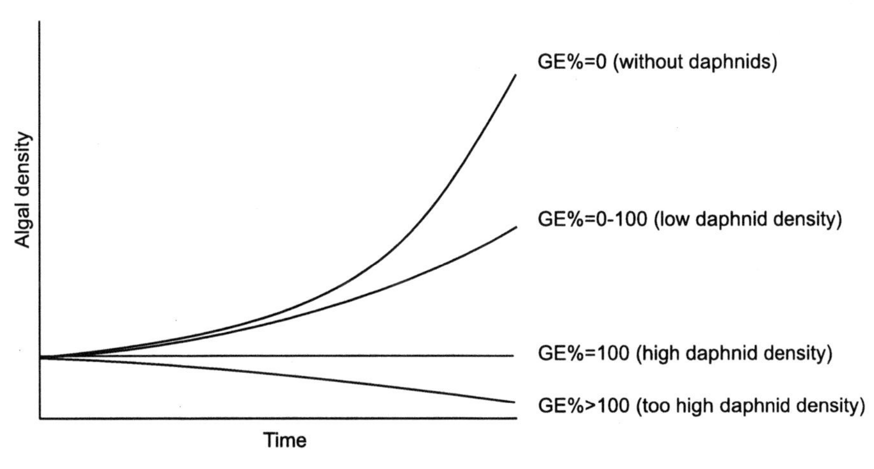

Fig. 2.10. Algal development under unlimited conditions in various scenarios with respect to daphnid grazing efficiency. In the absence of daphnids (no grazing) the exponential curve can be characterized by a relative growth rate "r". With daphnid grazing, a residual growth will be recorded that can be characterized by nett growth rate "r_g". A relative grazing effectiveness (GE%) for the daphnids can be calculated from "r" and "r_g"(see text). Grazing equals algal growth at a GE of 100%. When the daphnid grazing exceeds algal growth, a GE > 100% is recorded

How to Quantify Grazing Effectiveness

The daphnids' grazing effectiveness upon certain algae under various conditions can best be tested in an experimental situation in which algae intrinsically have an exponential growth in density. Comparison of the algal development in a situation with and without daphnids indicates the daphnid grazing effectiveness.

In a situation in which nutrient availability, light conditions or daphnid grazing do not limit algal growth, the intrinsic relative growth rate ('r') can be estimated when assuming exponential growth (see Fig. 2.10):

$$Chl_{(t=x)} = Chl_{(t=0)} \cdot e^{(r \cdot t)}$$

where: Chl = algal biomass (chlorophyll concentration)

t = time in days

r = intrinsic relative growth rate algae (day^{-1})

In a situation with daphnids, the algal density development will be reduced due to grazing. In this case the residual rate of algal development ('r_g') can be estimated from the recorded growth curve (see Fig. 2.9).

Comparable to the Algal Response Factor used by Sarnelle (1992), a grazing effectiveness factor (GE) can be calculated as:.

$$GE = r - r_g$$

The relative grazing effectiveness can be expressed as the percentage reduction of the algal growth development (GE%):

$$GE\% = GE/r = (r - r_g)/\ r \times 100$$

In the absence of daphnids GE% will be zero, while with sufficient daphnid density the algal development will be completely controlled, which will be expressed as GE%=100. GE% can reach values above 100 in cases where the algal consumption exceeds the production due to poor algal development and / or very high daphnid densities (Fig. 2.10).

A Plankton Eco-assay for Measuring Daphnid Grazing Effectiveness

The plankton eco-assay was developed in order to generate data that can be used to calculate grazing effectiveness. The plankton eco-assay is an indoor microcosm test suitable for studying algal development under controlled conditions with or without daphnids. More information on the set-up of plankton eco-assays is given in Sect. 5.2. In our studies we used 80-litre test systems, with a hyper-trophic medium stocked with cultured algae. The systems were continuously aerated and illuminated with a day-night regime (Fig. 2.11). Daphnids were added to part of the systems in order to quantify the grazing effectiveness from a comparison of algal growth in systems with (Rg) and without (R) daphnids over a period of 5–8 days. Daphnid density counting at the end of the test provided information on both the mortality and the reproduction of the added daphnids.

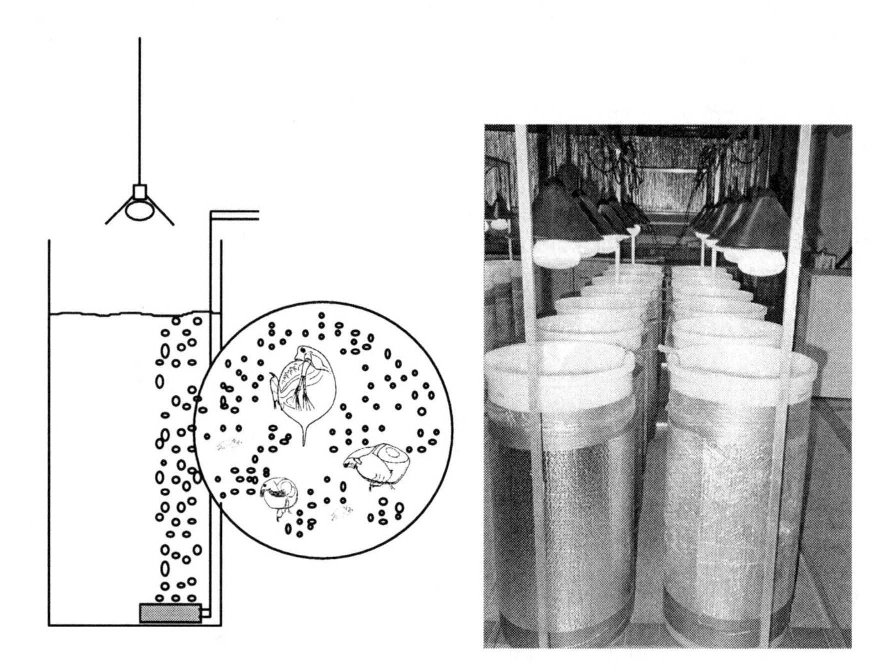

Fig. 2.11. The plankton eco-assay: a test system for determination of the daphnid grazing effectiveness under various circumstances

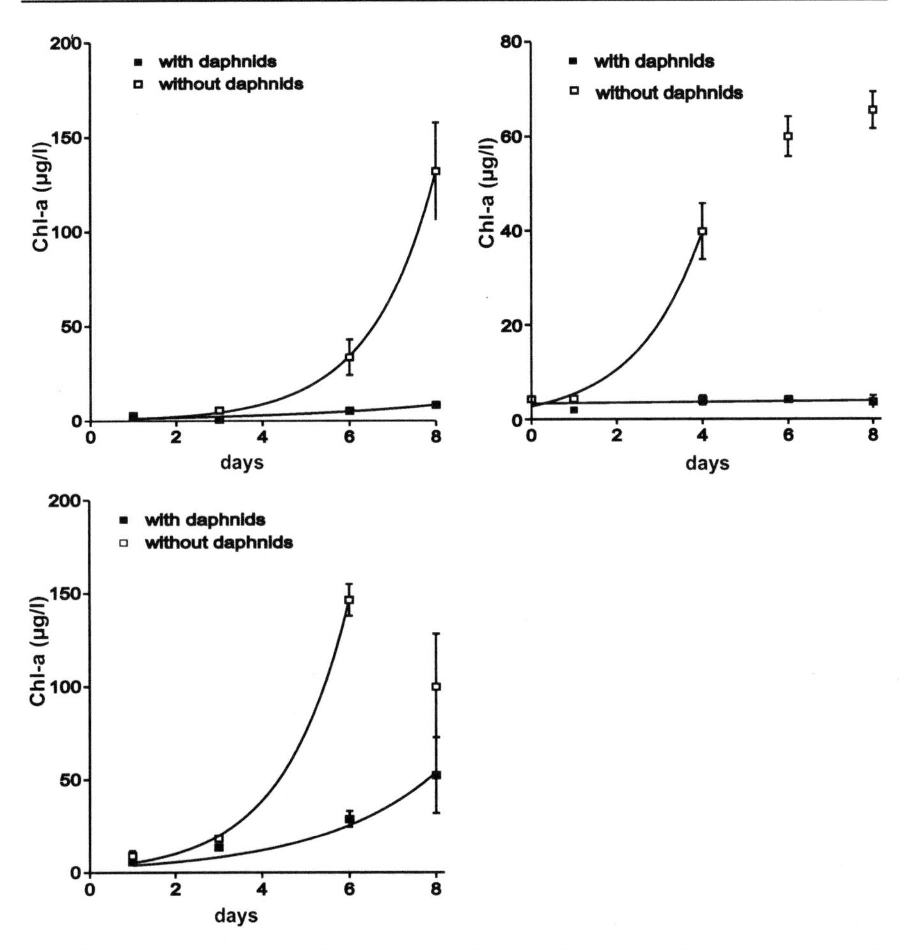

Fig. 2.12. Algal development in eco-assays with and without daphnids in situations with exponential development throughout the test (top, left), limited algal development (top, right) and collapsing algal densities (bottom, left). Algal development rates R were calculated for the period with 'exponential growth: 0.67, 0.67 and 0.66 d^{-1} respectively. The corresponding grazing effectiveness GE% was 49%, 88% and 39% respectively

A typical example of the development of algae in an eco-assay is presented in Fig. 2.12 (left). This test was performed in OECD algal medium stocked with *Chlorella pyrenoidosa* at an initial chlorophyll-a concentration of approximately 5 µg/l. Two of the four test containers were stocked with an assemblage of daphnid species (*Ceriodaphnia* sp., *Daphnia magna* and *Simocephalus vetulus* at 3, 2 and 3 individuals, respectively, per litre). The two other containers without daphnids were used as a reference. The algal population in the containers without daphnids showed an exponential development with a growth rate R of 0.67.d^{-1}. In the daphnid-containing containers, the algal development Rg was substantially reduced to 0.34.d^{-1} (corresponding to a grazing effectiveness GE of 0.33 d^{-1} or GE%=49%) and was almost completely controlled by the daphnids. It must be stressed that the grazing effectiveness was measured in relation to the control of an algal population that was reproducing at an exponential rate. Whilst a grazing efficiency of 49% may sound modest, after eight days the algal density in the systems with daphnids was only 6% of that in the systems without daphnids.

In most of the tests the algal development continued during the whole experimental period (Fig. 2.12, top, left). However, this development occasionally became limited after a few days (Fig. 2.12, top, right) or even collapsed (Fig. 2.12, bottom, left). In these cases, the data used for the calculation of the algal development rate were restricted to the period of exponential algal growth.

Daphnid Grazing Effectiveness in Response to Algal Composition

Different growth rates were calculated for the various algal species used in the eco-assay (Fig. 2.13). The highest growth rates were found for the green algae *Raphidocelis subcapitata* and *Scenedesmus subspicatus,* both being approximately 0.7 per day, which is significantly higher than the growth rate of the green algae *Chlorella pyrenoidosa*. In a test that was performed with the cyanobacteria *Oscillatoria sp.*, a growth rate of 0.36 per day was calculated which, as expected, is low in comparison to that of the green algae. A low growth rate is a general characteristic of cyanobacteria (Schreurs 1992; Andersen 1997).

The daphnid grazing effectiveness indicated remarkable differences in the palatability of the various algae. The small spherical algae *Chlorella pyrenoidosa* was readily and effectively grazed (average GE 0.31d^{-1}; average GE% 57). The slightly larger and more elongated species *Scenedesmus subspicatus* was consumed at nearly the same rate as *Chlorella* (GE 0.28 d^{-1}), but due to its higher growth rate, the average grazing effectiveness was only 43%. The third green algae tested,

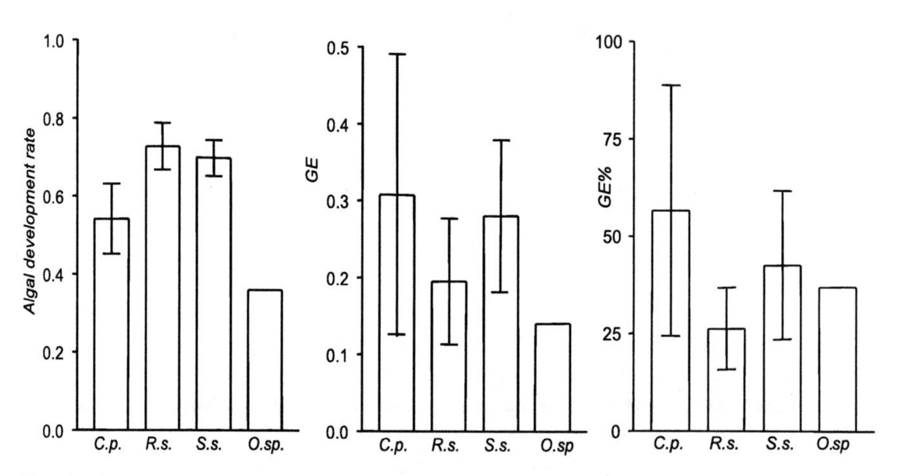

Fig. 2.13. Relative growth rates of different algal species as determined in plankton eco-assays without daphnids, as well as daphnid grazing effectiveness (GE and GE%) on those algae in simultaneous plankton eco-assays with 10 daphnids per litre. Initial algal density was ca. 5 µg/l chlorophyll-a. Average values are presented with standard deviations for *Chlorella pyrenoidosa* (C.p. n= 9), *Raphidocelis subcapitata* (R.s. n=4), *Scenedesmus subspicatus* (S.s. n=7) and *Oscillatoria sp.* (O.sp. n=1)

Raphidocelis, was grazed with the lowest effectiveness (average GE% 26), due to its high growth rate and the low grazing rate (GE 0.20 d^{-1}). The low palatability of *Raphidocelis* was probably the result of its curved shape, making it a relatively difficult species to handle. In response to environmental stress (such as nutrient limitation, grazing pressure, etc.) *Raphidocelis*, like many other algal species, may form large filaments or other anti-grazing defences, which may reduce cladoceran grazing efficiency (Van Donk and Hessen 1993; Van Donk et al. 1997).

The cyanobacteria *Oscillatoria*, was only marginally predated upon (GE = 0.13 d^{-1}), but due to its low exponential growth rate (R=0.36 d^{-1}), the grazing effectiveness (GE% 37) was comparable to *Raphidocelis* and *Scenedesmus*. Daphnids are known to have problems when it comes to the grazing of filamentous algae such as the cyanobacteria *Oscillatoria*. Although some daphnids are able to ingest *Oscillatoria* filaments up to 1 mm (Lampert 1987c), long filaments generally interfere with food processing efficiency, causing a higher energy expenditure and resulting in a lower nett growth rate (Repka 1996 1997; Rohrlack et al. 1999). As a result, relatively high cladoceran densities are needed to control *Oscillatoria*, even though this is a slow growing species (see above). A density of 9 ind/l *D. magna* can initially control *Oscillatoria*, but after a few days the algae 'escaped' control and its density rapidly increased. Like many other algae, *Oscillatoria* is probably able to respond to herbivore grazing by forming larger filamentous aggregates (Lampert 1987b; Hessen and Van Donk 1993; Zurek and Bucka 1994). This may further reduce palatability and, consequently, grazing efficiency, thereby explaining the time lag observed in the 'escape' from grazing control. A *Daphnia* density of 18 ind/l results in a grazing pressure that can prevent such "escape" responses.

It was also observed, in an eco-assay conducted with *Anabaena* sp., that the grazing effectiveness on filamentous cyanobacteria is heavily dependent on the initial daphnid densities. (Fig. 2.14). The growth rate of *Anabaena* was 0.51 d^{-1} and the grazing effectiveness of *Daphnia magna* was only marginal (GE% 2–5) at initial densities of 4 or 6 ind/l. A grazing effectiveness of 20–50% was obtained at initial densities of 8–12 ind/l , resulting in moderate control of the *Anabaena* population development.

Although green algae have a higher growth rate than cyanobacteria, they can be controlled by a lower number of daphnids. An initial daphnid density of between 5 and 10 individuals (*D. magna*, *D. longispina*, *Ceriodaphnia* sp.) was sufficient to control the development of *Chlorella pyrenoidosa* from initial chlorophyll-a density up to 15 µg/l chlorophyll-a in various eco-assay tests. This control failed at higher initial algal densities, because the number of algal cells per individual daphnid was too high.

For this reason, the control of the development of faster growing algal species, such as *Raphidocelis* and *Scenedesmus*, requires relatively higher daphnid densities. With *Scenedesmus* (Fig. 2.15), a density of 7 daphnids per litre resulted in a grazing effectiveness of only 33 %, and was therefore insufficient for control of the algal development. At higher daphnid densities (13 and 20 ind/l) the algal development was more or less controlled with a grazing effectiveness of 60–80%. The highest grazing effectiveness with *Raphidocelis* in a plankton eco-assay was reached with initial densities of at least 10 daphnids per litre, but was still only 56%.

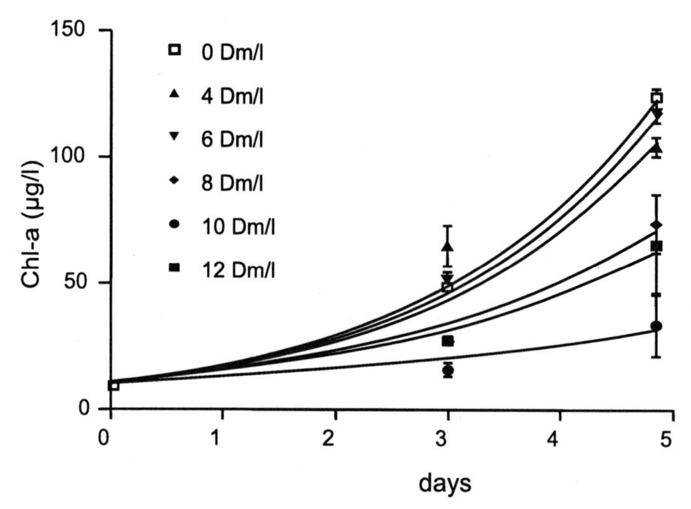

Fig. 2.14. Density development of the cyanobacteria *Anabaena sp.* in systems with different initial densities of *Daphnia magna* (Dm)

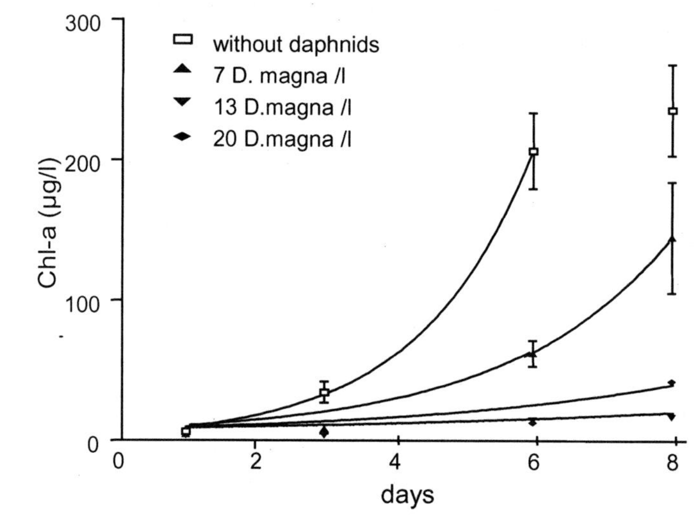

Fig. 2.15. Development in an eco-assay with OECD medium stocked with *Scenedesmus subspicatus* and various densities of *Daphnia magna*

The daphnid grazing effectiveness as a function of the daphnid densities in the control of *Raphidocelis* development was tested in more detail in smaller-scale microcosms (2 litre), which, due to their size provided an almost complete control of the experimental conditions. The added daphnids consisted of individuals of similar age and size, and were monitored by visual count throughout the test. It appeared that the development of *Raphidocelis* could be completely controlled in these test systems by 20 daphnids per litre (Fig. 2.16).

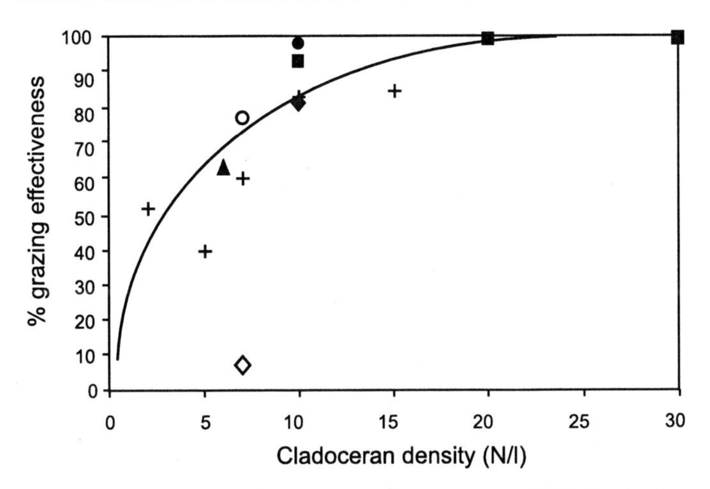

Fig. 2.16. Daphnid grazing effectiveness versus daphnid density in seven 2L eco-assays. Each symbol reflects an individual experiment

Daphnid Grazing Effectiveness of Various Daphnids

The variation in daphnid grazing effectiveness in the tests presented is partly caused by the differences in the tests with respect to the composition of the daphnid communities used.

In experiments with both *Chlorella* and *Raphidocelis*, *D. magna* emerged as the most effective and *S. vetulus* as the least effective grazer, with *D. longispina* in an intermediate position. In another test with *Chlorella* , however, *D. longispina* again demonstrated a better grazing performance than *S. vetulus,* but *D. Magna* failed. This might have been related to differences in the state of the *D. magna* populations.

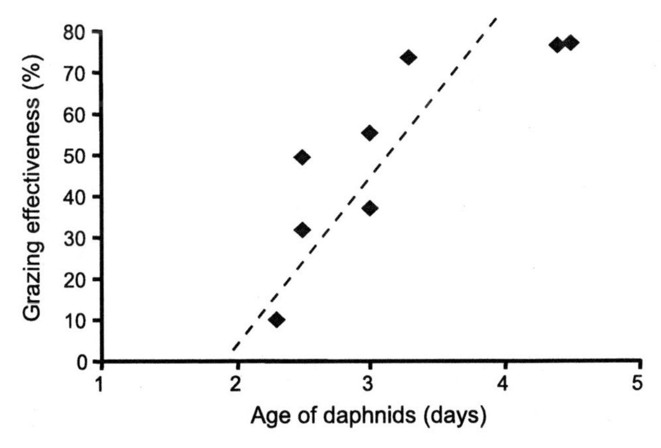

Fig. 2.17. Effect of mean daphnid age (start of the experiment) on the grazing effectiveness. Experiment carried out in 2-litre microcosms

From a 2 litre microcosm test, it appeared that the impact of daphnids on the algal development heavily depended on their size (i.c. age). Seven daphnids of at least 3 days old were capable of a grazing effectiveness of more than 50% with the algae *Raphidocelis,* while at least 14 neonates (age < 1 day) per litre were required to achieve the same effectiveness (Fig. 2.17).

The grazing effectiveness of individual daphnid species was also compared with the grazing effectiveness of a multi-species daphnid population. A weak verification of the hypothesis that a multi-species combination might be more effective in grazing than a monospecies community, was confirmed in only one of the tests.

Population Response

The plankton eco-assay is a useful tool for measuring grazing effectiveness. It provides information on the short-term response (grazing effectiveness) of daphnids to algal growth. However, because of the short time scale and the simple community level in the eco-assay, these results should be used very carefully for longer-term field predictions. Enlarging the test system and the test period reduces the uncertainty for field predictions, but makes the evaluation of the results more complex. The microcosm size required depends on the experimental duration desired in regard of the life cycle of the organisms to be tested (Cairns and Cherry 1993) (Fig. 2.18).

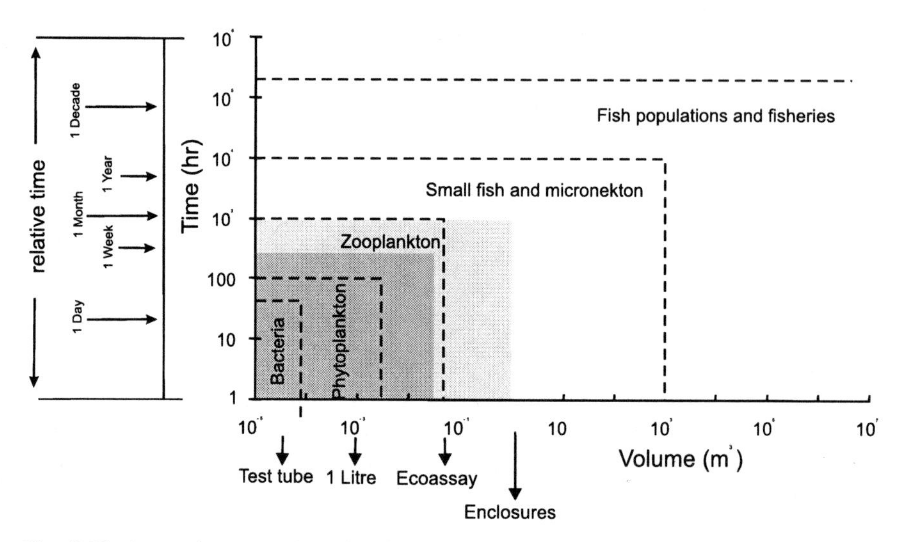

Fig. 2.18. Approximate relationship between aquatic organism life cycles and the relative size of their containment (modified from Parsons 1982). The shaded area indicates the size and duration of the enclosures used in this study

Plankton Enclosure Study

The 80-litre plankton eco-assay is an appropriate tool for testing the effect of daphnid grazing on algal communities at the simple community level over a period of ca. one week. When the effects of (reduced) grazing over the longer term must be determined, a larger scale test system and a longer experimental period is required. For this purpose, 1000–3000 litre outdoor plankton enclosures are employed (Fig. 2.19).

One of the factors that should be taken into account in larger systems (longer term studies) is the temporal dynamics of the daphnid population and the consequences for algal grazing. However, the results of these enclosure experiments will, when interpreted in the correct way, provide more reliable information on phytoplankton – zooplankton interactions in the field situation.

The fact that it is more difficult to set-up and evaluate the more complex experiments with the enclosure systems is also reflected by the practical impossibility of creating favourable initial conditions i.e. balancing the algal density with the daphnid density. This is in contrast to the relatively small experimental systems, such as the plankton eco-assay, where manipulation of the test conditions is possible. In the enclosures, the initial daphnid density was too low to control the algal development from the start, especially when this algal development was stimulated by nutrient additions.

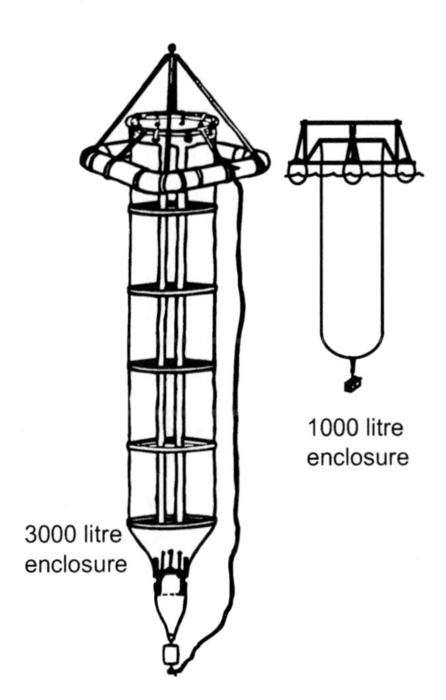

Fig. 2.19. Outdoor plankton enclosures used for the experimental studies on plankton dynamics at community level

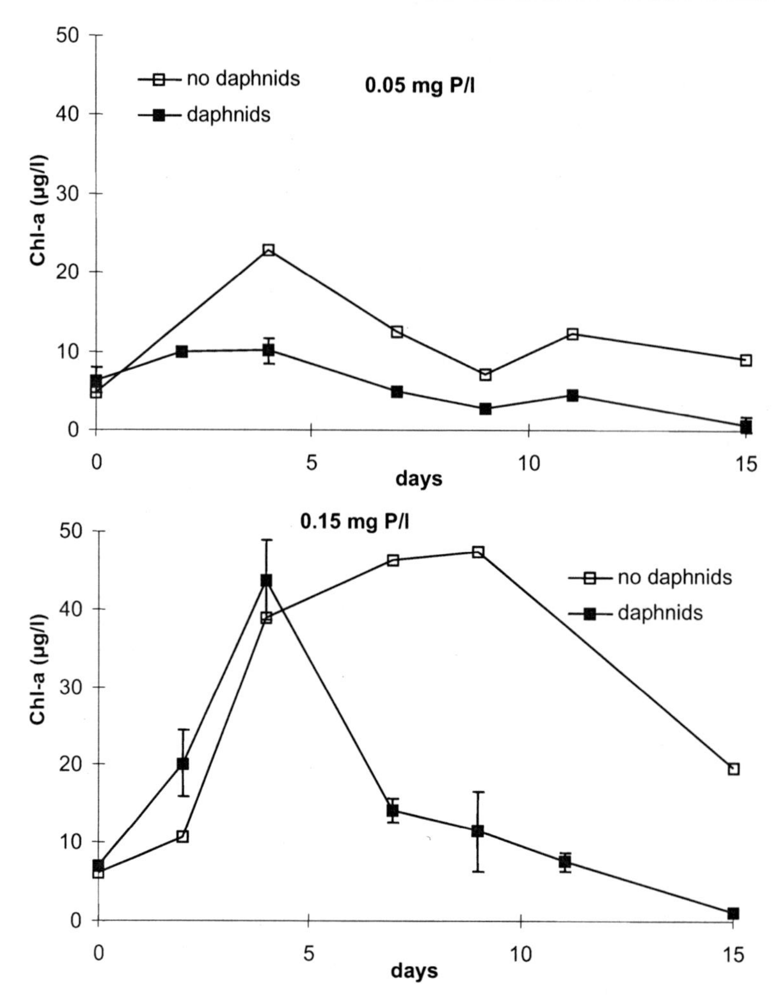

Fig. 2.20. Algal development in 3000 litre plankton enclosures with and without daphnids using two different nutrient loads. No daphnids: sieved over a 150 µm sieve. P:N ratio was 1:22

The relatively low initial daphnid densities allowed the response of the daphnid population to enhanced algal development to be studied. It was observed in various tests that the daphnid community needed some time to reach densities sufficiently high to control the algal development.

As an example, the results of an experiment that was performed in the 3000 L enclosures are presented (Fig. 2.20).

The experiment showed that control of the algal development was reached sooner in the systems with low nutrient loading. This is not surprising, since the algal development at low nutrient loading could already have been controlled by a *Daphnia longispina* density of ca. 1 individual per litre, while it took more than 3 individuals per litre to control the algal development in the enclosure with high nutrient loading. The faster the growth rate of the daphnid community, the better it will be able to respond to potential algal blooms. The reduction in the algal density at the end of the experimental period was due to nutrient depletion.

Daphnid populations apparently respond to increased primary production by rapid reproduction, ultimately resulting in the re-establishment of top-down control. However, in high-nutrient, high-productivity systems, this may require a response time of several days. In the longer term, the daphnid density and the related grazing rate adapt to the algal growth rate, resulting in an algal concentration around the critical food level of the daphnid community.

Synopsis

It has been clearly demonstrated that daphnids are capable of controlling the development of algal density by grazing. The grazing effectiveness of the daphnid community depends on various factors. The algal species plays an important role and fast growing or less edible (such as filamentous cyanobacteria) algae are more difficult to control. However, in most cases even these algal species can be controlled by grazing, as long as the daphnid density is high enough. It thus appears that daphnid density, in relation to the algal growth rate (depending on algal composition, algal density and nutrient availability) and the algal palatability, is the critical factor in the effectiveness of the grazing response of daphnids to algal growth. The "Potential Grazing Pressure" (ratio between daphnids and algae), as introduced by Matveev and Matveeva (1997), is therefore a relevant indicator for potential grazing effectiveness by daphnids, as long as differences in daphnid and algal composition are considered.

2.3 Modelling Daphnid Grazing Effectiveness

Calculating Grazing Effectiveness from Ecoassays

The eco-assay is used to give a quick insight into the processes of grazing and algae growth, with the maximum algal growth rate (r) and the grazing effectiveness GE (or GE%) as steering parameters. In the interpretation of the eco-assay results (i.e. the calculation of the grazing effectiveness), GE is taken as a fixed parameter over the experimental period. This is a useful simplification that holds for the interpretation of the eco-assay results, but the question remains as to what impact grazing effectiveness, or ineffectiveness, can have on the algal and dahpnid development over the longer term i.e. the seasonal plankton development with a time variable GE.

For this purpose, an 'as simple as possible' model was developed that describes the development of algal and daphnid densities over time. It was not intended to model a complete complex plankton community.

In the aforementioned model, the change in daphnid density is specified by growth and mortality while the change in algal density is specified by growth and grazing. The grazing effectiveness of the daphnid community (that was assumed to be constant over the experimental period of the eco-assay) is now related to the

Table 2.3. Model equations for the interpretation of ecoassay results

$$\frac{d[daphnids]}{dt} = (r_d - m) \times [daphnids] \qquad \text{Eq. 1}$$

$$\frac{d[Chla]}{dt} = (r - GE) \times [Chla] \qquad \text{Eq. 2}$$

$$GE = CR \times [daphnids] \qquad \text{Eq. 3}$$

$$r_d = \frac{GE \times [Chla]}{[daphnids]} \times Cf \qquad \text{Eq. 4}$$

With,

[daphnids]	= daphnid density	ind/l	
[Chl a]	= algae concentration	µg/l	
r_d	= daphnid growth rate	d^{-1}	max. 0.3
M	= daphnid mortality rate	d^{-1}	const. 0.033
r	= algae growth rate	d^{-1}	
GE	= grazing effectiveness	d^{-1}	
CR	= daphnid clearance rate	$l.ind.^{-1}.d^{-1}$	
Cf	= conversion factor	$ind.µg[Chl\ a]^{-1}$	

daphnid density and the filtration capacity or clearance rate of the daphnids. The daphnids' growth rate , also assumed to be constant for the purposes of the eco-assay, is in fact a function of the grazed amount of algae per individual and a conversion factor, which indicates the effective use of grazed algae biomass by the daphnids for their reproduction. The model does not take into account complicating factors such as seasonal variation or fish predation which, for this purpose, is not restrictive. The model's equations are presented in Table 2.3. Apart from the mortality rate in Eq. 1, Eqs. 1 and 2 are the same as those used to calculate the grazing effectiveness GE in the interpretation of the eco-assay results.

The maximum growth rate for daphnids is set at 0.3 d^{-1}. This value was taken from literature and from experimental data as an average value for several daphnid species . Not only is the growth rate of the daphnids limited, but their clearance rate has a maximum value, which indicates the maximum volume of water filtered per individual daphnid per day. The conversion factor has a minimum value, which indicates the minimum grazed amount per daphnid from which sufficient energy is obtained for the continuation of reproduction. When the model is fitted onto experimental data, the values of the clearance rate and the conversion factor should be realistic.

The model does not include the minimum or maximum algal densities that exist in the field situation and that are caused by predation, nutrient limits and other limiting factors. However, it describes the development of the algal and daphnid densities in non-limited systems, with and without daphnids, in an acceptable way.

Calibration from Experimental Data

The model was fitted on the results of two plankton eco assays performed with two algal species with different growth rates and palatability: *Chlorella pyrenoidosa* and *Raphidocelis subcapitata* (Fig. 2.21).

The results show that daphnids' growth responded rapidly to the growth of the *Chlorella* population, resulting in an almost immediate control of the development of the algal density. *Chlorella* was efficiently grazed and assimilated by the daphnids. After one week, the algal density became limiting for the daphnid development and daphnid densities began to decline. The maximum daphnid density that was reached in the *Chlorella* systems was ca. 50 individuals per litre.

The other algal species, *Raphidocelis*, had a much higher growth rate than *Chlorella*. Although this lead to high algal availability for the daphnids in this test, it was not reflected by a strong development of the daphnid density. *Raphidocelis* was apparently not as efficiently grazed upon as *Chlorella*. This observation is represented by the relatively low conversion factor for *Raphidocelis*. Because of this low grazing effectiveness, in combination with the high algal growth rate, the daphnid population could not respond with sufficient speed to prevent the development of high algal densities. Nonetheless, the impact of daphnid grazing on the algal development was still clear, and after 9 days the daphnid density was sufficiently high as to prevent the further development of the algal density. At that moment almost 100 daphnids per litre were present. The model fitted nicely with the experimental results and showed the difference between the two algal species.

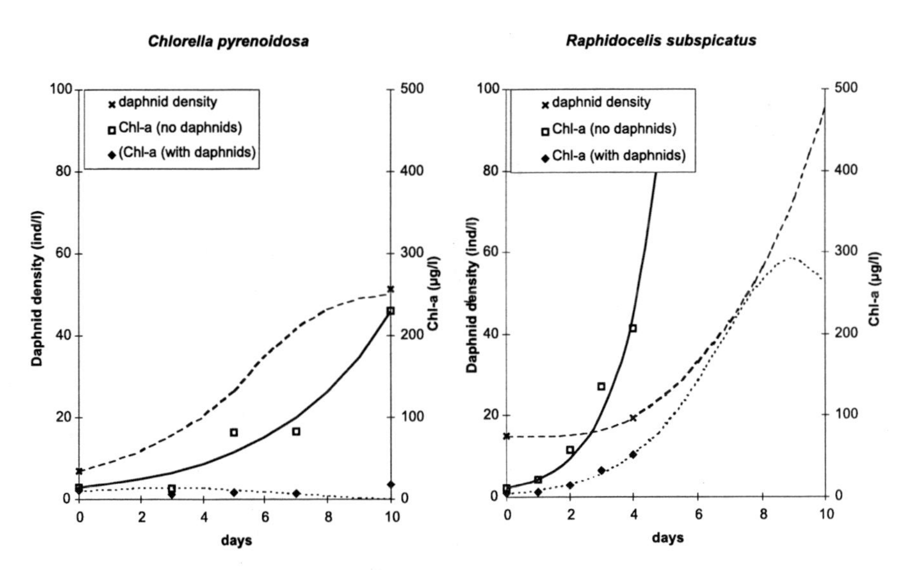

Fig. 2.21. Experimental data (observations) from planktoneco assays with two different algal species fitted with the simple grazing model

Daphnid Density Effect

It is known from the eco-assay experiments that the initial daphnid density is one of the main parameters determining the control of the algal biomass in experimental systems. In order to analyse the impact of the initial daphnia density, the model was run for several daphnid density settings. The parameter values of Cf and CR that were calculated during the fitting of the experiment with *Chlorella* were used in these (and further) exercises. Model runs with initial daphnid densities ranging from 0.2 up to 10 individuals per litre and a constant initial algal density confirmed this observation (Fig. 2.22). The maximum chlorophyll-a concentrations that were reached depended heavily on the initial daphnid density. The lower the initial daphnid density, the longer it took the daphnids to reach sufficiently high numbers to control the algal development, while a rapid response by the daphnid population prevented the development of high algal densities.

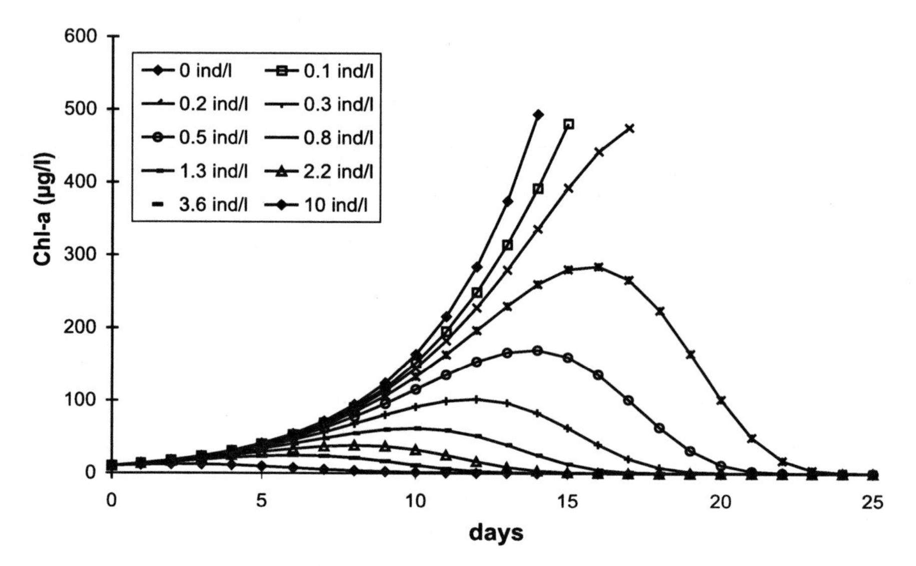

Fig. 2.22. Predicted development of the algal density (in Chl-a) as a function of time and initial daphnid density (expressed as individuals per litre)

It seems obvious that the initial algal density also played an important role in the possibility that daphnids may also control the algal development. However, model runs with various initial algal densities and fixed initial daphnid densities did not confirm this idea (Fig. 2.23). The initial density hardly impacted the nett algal production (Table 2.5). This can be explained by the fact that the daphnid development was faster at higher food availability and therefore compensated for the larger amount of algae grazed. Regardless of the initial algal density, the maximum algal and daphnid densities were reached on days 3 and 11, respectively.

Table 2.4. Predicted maximum algal (Chl-a) and daphnid densities and the day that the maximum daphnid densities were reached as a function of the initial daphnid density

Initial daphnids ind /l	max. Chl-a μg/l	max. daphnids ind/l	max. daphnids reached at day
0.1	494	76	>day 25
0.2	481	122	>day 25
0.3	475	116	Day 24
0.5	285	105	Day 22
0.8	170	94	Day 19
1.3	102	84	Day 17
2.2	62	74	Day 15
3.6	38	63	Day 13
6.0	25	54	Day 11
10.0	12	46	Day 10

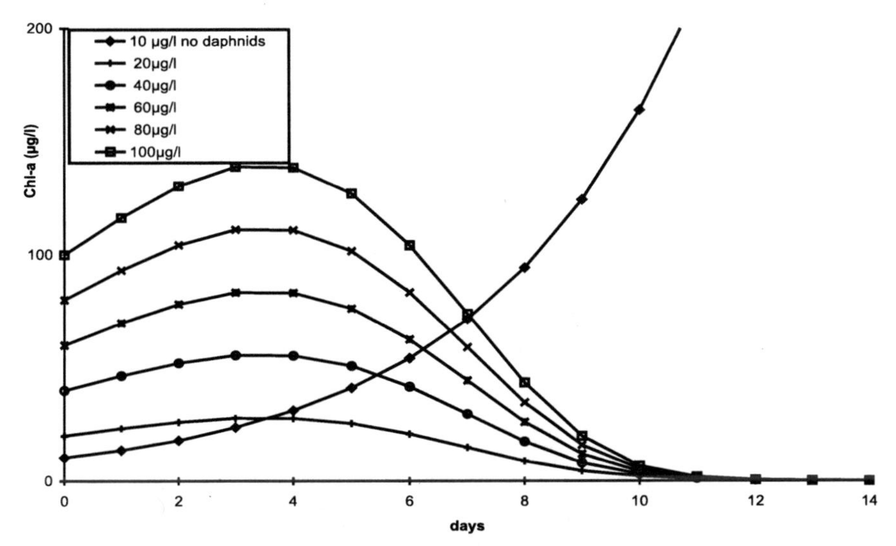

Fig. 2.23. Predicted development of the algal density (expressed as Chl-a) as a function over time, at different initial algae concentrations

However, the final daphnid density that was required to control the algal development was higher at higher initial algal densities. It took 66 daphnids per litre to control the algal development with an initial chlorophyll concentration of 20 μg/l and 101 daphnids per litre with an initial chlorophyll concentration of 100 μg/l. Moreover, it should be kept in mind that this model run was performed with *Chlorella* and daphnids showing a high grazing effectiveness. If the same exercise were to be performed with a lower grazing effectiveness, the daphnid population would be unable to respond with similar efficiency, the impact of the initial algal density would be more pronounced and even higher daphnid densities would be necessary to stop the algal growth.

Table 2.5. Maximum daphnid and algal densities and nett algal production (calculated as maximum – initial Chl-a) in relation to the initial algal concentrations. In all cases, maximum values were reached on days 3 and 11 for algae and daphnids respectively

initial Chl-a µg/l	max Chl-a µg/l	nett Chl-a production µg/l	max daphnids ind/l
20	28	8	66
40	56	16	82
60	83	23	91
80	111	31	97
100	139	39	101

Effect of Reduced Grazing Effectiveness

Finally, the impact of the grazing effectiveness (GE%; as measured from an ecoassay experiment) on the algal development was determined with the model (Fig. 2.24). A reduction in the grazing effectiveness can be caused by factors unfavourable to the daphnids, such as low palatability of the algae or a poor water quality. Simulation of a reduced grazing effectiveness, represented by reduced values of CR and Cf, resulted in a delayed response by the daphnid population and therefore in higher maximum algal densities. The situation is comparable with what was predicted as necessary for algal control at both low initial and higher daphnid densities. Due to the slow response by the daphnids, the algal community reached high densities that required even higher daphnid densities for control. When the grazing effectiveness was significantly reduced (GE ≤ 10%), the daphnids were no longer able to control the algal development, and the result was a prolific growth of algae.

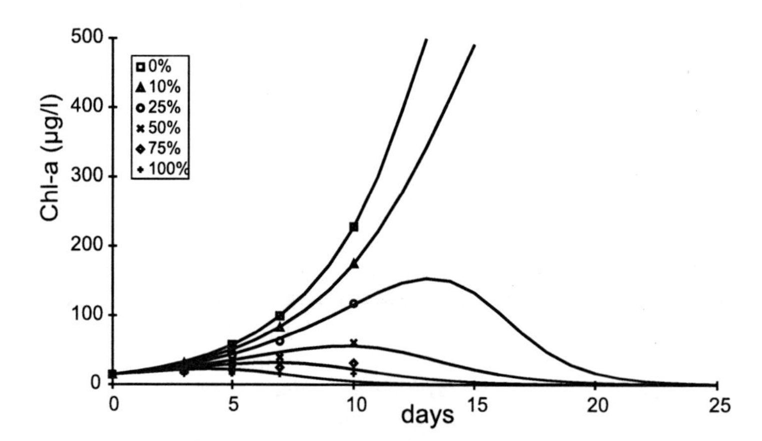

Fig. 2.24. Predicted development of the algal density in relation to the daphnid grazing effectiveness GE. The relative grazing effectiveness (legend) is presented as a percentage of the actual GE determined in the plankton ecoassay

Table 2.6. Maximum algal (Chl-a) and daphnid densities and the day on which these maxima were reached, based on model runs with various grazing effectiveness (GE)

GE	max Chl-a (μg/l)	on day	max daphnids (ind/l)	on day
0%	500	13		
10%	343	15	497	25
25%	154	13	109	20
50%	56	10	56	18
75%	32	7	47	15
100%	22	4	50	11

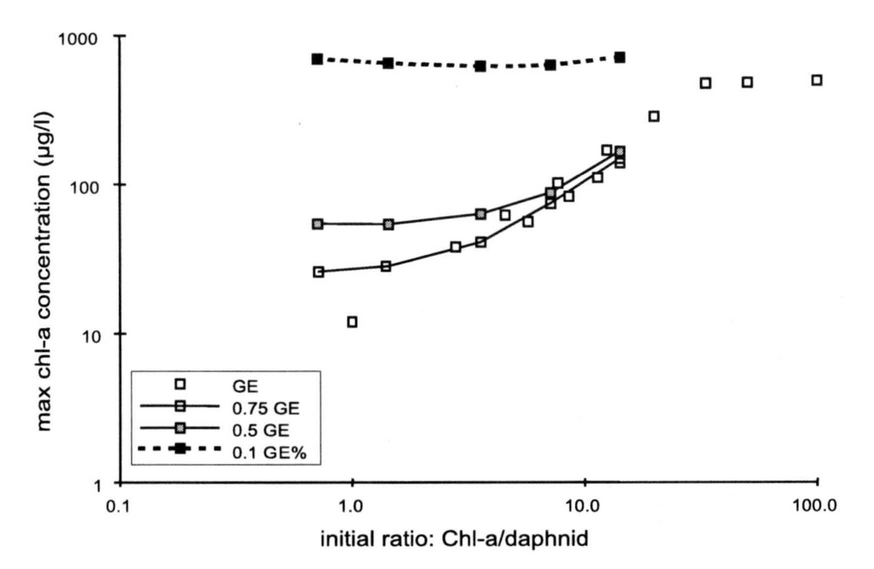

Fig. 2.25. Relation between the highest algal density (expressed as max Chl-a) and the initial Chl-a daphnid ratio, calculated in the model runs with various grazing effectiveness (GE)

What Can Be Learned from the Model Simulations?

It can be concluded from the model simulations that the maximum algal density that will be reached in an eco-assay with daphnids depends on three factors: the initial daphnid density, the initial algal density and the grazing effectiveness.

When the grazing effectiveness is constant, the maximum algal density correlates with the initial quantity of algae (chlorophyll) per daphnid (Fig. 2.25). When the grazing effectiveness is reduced, the capability of the daphnids to prohibit the development of high algal densities is severely reduced.

The model used was obviously much too simple for predicting plankton dynamics in the field situation. In the model, the daphnid density was not limited and in a model run all algal blooms will be finally controlled when the daphnids reach extreme numbers. In the field however, the circumstances for daphnids are

extreme numbers. In the field however, the circumstances for daphnids are less favourable. Mortality rates can be high due to predation and the maximum density is always limited by environmental conditions such as fish predation, food and oxygen availability or habitat characteristics.
Under natural circumstances where the maximum daphnid density is limited, a rapid response by the daphnid community to increasing primary production is essential. If this response is delayed, the algal community can reach high densities, which are beyond the control of even the maximum daphnid density. The further development of the algal density in this situation then almost completely depends on nutrient availability and light conditions. The result is enhanced algal biomass densities, frequently shifted to less edible algae.

An accurate response by the daphnid community in a critical period is therefore essential for the prevention of lasting algal blooms that might switch a clear water state into a turbid eutrophied water state. Since the exponential growth of phytoplankton is strongest during a seasonal bloom, the most important period affecting the daphnid response is early in the period of algal bloom (i.e. spring in temperate regions).

3 Toxic Reduction of Daphnid Grazing Effectiveness

3.1 Daphnid Ecotoxicology

Daphnids as Ecotoxicological Test Species

During the past decades, a large number of freshwater invertebrates have been used in ecotoxicity testing. However, world-wide, only the cladoceran crustaceans have emerged as a key group for standardised ecotoxicological tests (Persoone and Janssen 1993). Scientific as well as practical considerations directed the choice of cladocerans (Baudo 1987; Persoone and Janssen 1993; Mark and Solbé 1998):

- Cladocerans occur world-wide in a large variety of freshwater types;
- They fulfil a key role in freshwater food chains;
- They are assumed to be relatively sensitive compared with other organisms;
- They are sensitive to a wide variety of contaminants;
- They are relatively easy to culture in the laboratory;
- Reproduction is normally parthenogenic;
- They have a short life cycle;
- They can be tested in small volume test systems.

The standardised tests that have been developed assess the relative toxicity of existing and newly developed chemical substances. A tiered approach is usually followed in environmental hazard and risk assessment. A first screening is performed consisting of simple short duration tests, which identify clear end-points, such as mortality. If this first screening indicates that toxic effects are expected, more sophisticated tests are set up to determine any sub-lethal endpoints resulting from chronic exposure. Tests are usually performed on species from three major taxa: algae, crustaceans (and/or insects), and fish. For pesticides, this tiered approach, which is required in regulatory procedures, may eventually result in the execution of mesocosm experiments in order to assess the environmental effects in test systems that mimic field conditions, including field relevant exposure and biodiversity.

The cladoceran species most frequently used are *Daphnia magna, D. pulex* and *Ceriodaphnia dubia* (OECD 1993a), but many other cladoceran species have been used in ecotoxicological testing (Baudo 1987). In general, the sensitivity variation between cladoceran species is less that the variation between invertebrate species. In selecting cladocerans as the standard, the OECD regarded overall sensitivity and widespread experience with the test species as being of more importance than the testing of local species which might be relevant to (local) waters giving cause for concern (OECD 1993a).

Daphnia magna and *D. pulex* are the most commonly used species in acute toxicity tests for the purpose of assessing the relative toxicity of chemical substances in which mortality is used as the toxic end-point. In other words, the number of individuals killed by a range of substance concentrations in water is recorded after an incubation period of 24 or 48 hours. Several ring tests have shown that the test has a reasonably well intra- and inter-laboratory repeatability (Persoone and Janssen 1993).

Both *Daphnia* and *Ceriodaphnia* species are also used for chronic tests in which survival and reproduction (number of young produced) is recorded over a period of 21 and 7 days, respectively. Water is renewed daily or three times a week, and the animals are fed daily. As in acute tests, exposure to the test substance is exclusively via the water since the test set-up does not facilitate uptake via the food.

Daphnids and other cladocerans appear to be very sensitive to various pesticides, most notably insecticides. In Table 3.1, the toxicity of pesticides to cladocerans, algae and fish is listed, showing the sensitivity of cladocerans. The relative sensitivity of cladoceran species compared to other invertebrates and algae has also become evident from several mesocosm studies (see Sect. 3.2) in which the response of different populations can be observed resulting from the same toxic exposure concentration.

Standardised single-species ecotoxicological tests are especially useful when comparing chemicals (e.g., for notification purposes). When assessing the impact of substances in actual field situations, however, different information may be needed, including data with more chemical and biological 'realism'. The extrapolation of standardised test results to field situations for ecological risk assessment may be troublesome for several reasons (Baudo 1987; Baird et al. 1991):

- Laboratory tests indicate the sensitivity of a certain, inbred, laboratory population, which does not necessarily resemble that of genetically diverse field populations;
- Laboratory test conditions may be markedly different from field conditions;
- The test species is not always representative of the most sensitive member of a given community for a given contaminant in the field;
- Food-web interactions between species in an aquatic community are neglected;
- The bioavailability of the substance differs from the field (see below).

Role of Food

The role of (food) particles in the bioavailability of substances is usually ignored in toxicity testing. In standardised tests, single toxicants are tested to which the organisms are primarily exposed via the water. In acute tests, daphnids are not fed at all during exposure. This means that exposure and uptake of the contaminant only takes place via the water. In chronic tests, daphnids are fed, but exposure to the contaminant is still mainly via the water and not via the food, since the food is

Table 3.1. Ecotoxicity (left, EC_{50}; right, NOEC) of selected pesticides for several groups of aquatic organisms. Figures in italics indicate that Cladocerans were more sensitive than algae; bold figures indicate that cladocerans were the most sensitive group tested (adapted from Ordelman et al. 1993 a,b,c; 1994; and Teunissen-Ordelman et al. 1995 a,b,c; 1996 a,b; Teunissen-Ordelman and Schrap 1996). F= fungicide, I = insecticide, H = herbicide, G = general (broad spectrum) – = no data available

Pesticide		**EC_{50} (µg/l)**				**NOEC (µg/l)**			
	Mode	**Cladocera**	**Algae**	**insect**	**Fish**	**Cladocera**	**algae**	**Insect**	**fish**
Anilazin	F	*490*	600	–	85		<1000	–	9
azinfos-methyl	I	1.6	–	1.5	<0.3	***0.1***	1800	1.3	0.1
Benomyl	F	*640*	1400	7000	170		–	–	–
Bentazon	H	***64000***	280000	–	–		–	–	–
Bifenthrin	I	**0.11**	–	–	0.35		–	–	–
Carbofuran	I	**23**	–	56	80		3200	–	–
Chloorfenvinfos	I	**0.1**	–	0.7	23		<1000	–	–
Chloridazon	H	**180**	–	–	34000		730	–	–
Cyfluthrin	I	***0.14***	>10000	0.7	0.6		100	–	0.03
Cypermethrin	I		–	0.02	0.4	**0.0066**	–	–	0.03
Cyromazin	I	>92800	–	450	>89700	***310***	320000	–	14000
Diazinon	I	**0.522**	–	3	1.7	***0.2***	10000	–	<3.2
Dichloorvos	I	**0.066**	–	25	170		–	–	–
Dichloraniline									
Dimethoaat	I	*2500*	290000	43	4480	***29***	32000	–	100
Endosulfan	I	240	–	2.3	0.014	***27***	700	–	0.2
Esfenvaleraat	I	0.27	–		0.088	***0.01***	1.7	–	0.01
ETU	–	***26400***	6600000	–	7500000	18000	–	–	1000000
Fenpropathrin	I		>2000	0.27	1.8	***0.22***	>2000	–	–
Lindaan	I	***0.51***	4600	1	15		2000	2.2	2.9
Malathion	I	0.8	–	0.34	41	***0.002***	10000	0.17	4
Maneb	F	***2.4***	3200	>4900	340	>56	–	–	<18
mecoprop	H	420000	–	–	170000	***3300***	56000	–	–
methoxychloor	I	**0.78**	–	0.98	7		0.25	–	–
mevinfos	I	**0.16**	–	5	11		50	–	–
oxydemeton-methyl	I	**3.3**	–	35	4000		100000	–	–
parathion(-ethyl)	I	0.028	–	0.023	17.8	***0.08***	15	0.1	0.17
propoxur	I		–	13	1400	***400***	1000	–	400
tributyltinchloride	G	5.3							
tributyltinoxide	G					***0.56***	18		
trifenyitinhydroxide	F	**0.011**	–	–	15		–	–	–
trifluralin	H		–	1000	4.2	***2.4***	33	>4	1
Zineb	F	***970***	1800	–	7200	10	500	–	<0.32
Ziram	F	*140*	1200	100	75		–	–	<3.2

not loaded with the contaminant at the tested concentration. The concentration of the substance does not have time to form an equilibrium between the concentration taken up or bound to the food, and the concentration in the water. Consequently, this experimental set-up used in chronic toxicity testing does not account for contaminant uptake via the food.

In field conditions, organisms are exposed to numerous combinations of toxicants that may be available via the water or via food particles. The uptake route for substances depends on some chemical characteristics. In field conditions, many chemicals, especially hydrophobic organic contaminants (but also several metals) show a marked affinity for organic or sedimentary particles. When ingestion of food particles is the prime exposure route for these chemicals, testing only waterborne concentrations may underestimate the actual exposure (Muñoz et al. 1996; Fliedner 1997; Taylor et al. 1998). For other substances (e.g., metals), binding to particles might considerably reduce the bioavailability, and testing in the absence of food particles will consequently overestimate the effects that will occur in the field at the same total concentration level (e.g., Wang 1987; Winner et al. 1990; Meador 1991).

Effects on Feeding

As presented above, mortality and reproduction are the most commonly used endpoints in ecotoxicological tests with daphnids. By and large, this is because these endpoints are relatively easy to observe and the resultant LC_{50} or EC_{50} is a statistically sound parameter, and provides useful information on the relative toxicity of a substance. Mortality is, however, not among the most sensitive parameters (Flickinger et al. 1982). Death, preceded by other effects, which may manifest themselves in an earlier stage (though not necessarily), or at lower concentrations, is the final stage. These sublethal effects that precede mortality may be demonstrated by several endpoints, such as reduced reproduction, growth, and feeding. Feeding is a necessary prerequisite for maintenance, growth and reproduction, and should, therefore, be of special interest for ecotoxicological studies. Moreover, feeding has a direct influence on the condition of individuals, and, therefore, has an indirect effect on population dynamics and, ultimately, on community structure (Fig. 3.1). Reduced grazing at the individual level is reflected in population development and for interspecies competition at a community level and thus in the grazing effectiveness of whole plankton communities. Reduced algal grazing can, therefore, result in a reduced food chain production based on planktivorous production. The enhanced algal densities play a key role in the development of eutrophication problems such as toxic or nuisance algae, turbid waters, anoxic conditions, bad smell, midge plagues, etc.

Toxic Anorexia

Reduced daphnid grazing effectiveness, or reduced feeding rate, due to toxic stress can be considered to be 'toxic anorexia'. Anorexia literally means loss of appetite. When referring to human appetite, Anorexia nervosa is a denial of nourishment by an individual to his or herself due to an, often irrational, fear of becoming fat or to bring about, often unnecessary, weight loss. In the case of toxic anorexia, the loss of appetite is induced by the presence of one or more toxic substances. This effect does not necessarily involve the binding of the toxic substance to a receptor, which results in damage to either tissue or metabolic processes.

The effects of toxic substances on the feeding pattern of zooplankton, referring to daphnids in particular, can be studied in different ways. In the first place, effects on the filtering rate can be assessed from observations of the beat rate of the thoracic appendages used to collect food items (Gliwicz and Sieniawska 1986). Video observations have recently been used to assess the impact of fluoranthene on Daphnia feeding rates. However, the filtering rate is more often estimated from the clearance rate, i.e. the decrease in the amount of food particles in the medium in which daphnids are incubated for a certain period of time (Reading and Buikema 1980; Kersting and Van der Honing 1981; Flickinger et al. 1982; Fernández-Casalderrey et al. 1993; 1994; Taylor et al. 1998; Hartgers et al. 1999). The ingestion rate can also be assessed through direct measurements of the amount of food, or tracer for food, accumulated into the gut or body tissue after a short exposure time (Pott 1980).

Irrespective of how it is measured, feeding rate is a more sensitive parameter than mortality (see Table 3.2). The difference for endosulfan is only slight, but for lindane it is greater by almost two orders of magnitude. Selenium and cadmium, at non-lethal concentrations, have also been shown to reduce the filtration rate of *Daphnia pulex* and *D. magna* respectively, (Reading and Buikema 1980; Taylor et al. 1998).

Feeding rate seems to be a slightly more sensitive parameter than reproduction (Reading and Buikema 1980; Flickinger et al. 1982). Figure 3.1 demonstrates that the reproduction effort is related to the amount of food that is taken up. With prolonged exposure, reduced feeding will result in reduced growth and reproduction and eventually in increased mortality (Taylor et al. 1998). A reduced feeding rate may have direct consequences at the level of the population, especially in the field situation where competition and predation play an important role, (Day 1989).

Changes in the feeding behaviour of various types of aquatic organisms have been studied previously. The feeding activity of *Gammarus pulex* juveniles on eggs of *Artemia salina* were studied by Blockwell et al. (1998) after 96–240 h exposure to the toxicants copper, lindane and 3,4-DCA. Reductions in gammarid feeding were identified at 12.1 µg/l copper or 8.4 µg/l lindane and 918 µg/l 3,4-DCA. When competition for food by the isopod *Asellus aquaticus* was induced, a reduction in feeding activity by *G. pulex* was observed after 240 h exposure to a lindane concentration of 6.5 µg/l.

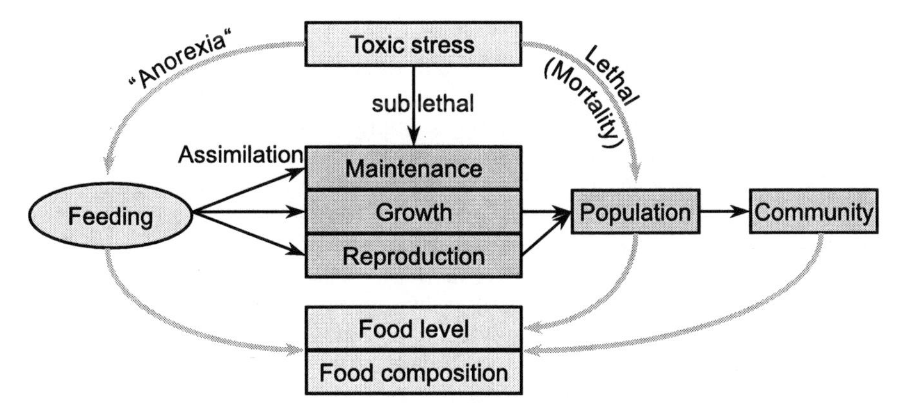

Fig. 3.1. Effects of toxic substances at individual, population and community level, and their interactions with food

Table 3.2. Effect concentrations of several toxicants for the feeding rate and mortality of *Daphnia magna* or *D. pulex*

Toxicant	Effect on feeding	Mortality	reference
Endosulfan	5hEC_{50}: 0.44 mg/l	24hLC_{50}: 0.62 mg/l	Fernández-Casalderrey et al. 1994
Diazinon	5hEC_{50}: 0.47 mg/l	24hLC_{50}: 0.9 mg/l	Fernández-Casalderrey et al. 1994
Methyl parathion	5hEC_{50}: 0.08 ng/l	48hLC_{50}: 0.31 ng/l	Fernández-Casalderrey et al. 1993
Atrazine	10minEC_{50}: 1.60 mg/l	48hLC_{50}: 9.88 mg/l	Pott 1980
Lindane	24hEC_{50}: 65 µg/l	48hLC_{50}: 516 µg/l	Hartgers et al. 1999
Dichlobenil	4hEC_{50}: 1 mg/l	48hLC_{50}: 7.8–10 mg/l[a]	Kersting and van der Honing 1981
Lindane	25% reduced: 0.05 mg/l	48hLC_{50}: 3.8 mg/l	Gliwicz and Sieniawska 1986
Copper	reduced at 10 µg/l	no mortality	Flickinger et al. 1982
Fluoranthene	reduced at 5 µg/l	48hLC_{50}: 8.7 µg/l	Taylor et al. 1998; Hatch and Burton 1999

[a] LC_{50} from Tooby (1978) referred to in Kersten and van de Honing (1981).

In conclusion, a reduced grazing rate or grazing efficiency may change the phytoplankton-zooplankton in the interactive plankton community dynamics. Reduced grazing by daphnids and its consequences for phytoplankton development have been studied in more detail in mesocosm experiments.

3.2 Observations from Mesocosm Studies

Experimental semi-field studies provide information on the impact of toxic substances on interactive biota at community level. These experimental systems include enclosures (in which part of a water column, sediment or part of a pond is enclosed between walls) and mesocosms (in which an entire ditch or pond system is imitated). The term 'mesocosm' is also often applied to enclosures.

Results from these types of mesocosm studies have shown that direct effects on zooplankton are often one of the most sensitive toxicity parameters, especially in the case of insecticides (deNoyelles et al. 1994; Foekema et al. 1996). As a result of the key role played by zooplankton in the grazing of phytoplankton, toxic effects on the zooplankton often result in indirect effects on other species, consequently affecting the functioning of the ecosystem as a whole (Hurlbert 1975; Jak and Scholten 1993). Thus, insecticide toxicity to the zooplankton may result in enhanced phytoplankton levels. In addition, increased phytoplankton levels may result in an increase in species that are competitors to zooplankton but which are less sensitive to the toxicant.

Some examples of the observed effects of toxic substances on the structure of zooplankton communities and their indirect effects upon phytoplankton development and benthic grazers (mediated by phytoplankton) are discussed below.

Effects of Toxic Substances on Plankton Communities

A series of enclosure studies was designed in order to test the validity of laboratory derived toxic effect concentrations for *Daphnia magna* under natural semi-conditions (Jak et al. 1996 and 1998). For this purpose, *D. magna* was cultured in the laboratory and stocked in natural water, with a native zooplankton and phytoplankton community from Lake IJsselmeer in the Netherlands. These experiments were carried out in 8 to 12 enclosures consisting of plastic bags suspended from buoyant frames (Fig. 2.19).

Dichloroaniline

The first experiment described here was aimed at validating the laboratory effect concentrations of 3,4-dichloroaniline (DCA) for *Daphnia magna* under natural conditions. Relatively high DCA concentrations (27, 90 and 270 μg/l) were applied. A single dose of phosphate and nitrate was added at the start of the experiment, in order to achieve the growth of algae as a food source for the zooplankton (Jak et al. 1998). A direct positive correlation was observed between phytoplankton densities, expressed as chlorophyll-a concentrations, and the level of DCA. This was due to a toxic reduction of *D. magna* population development (Fig. 3.2). The cyanobacteria species *Microcystus aeruginosa* dominated the plankton at the two highest concentrations.

Metals

In another experiment, the additive effect of a mixture of metals on a laboratory strain of *D. magna* was studied (Jak et al. 1996). The eight metals were added in equitoxic concentrations (i.e., in the same number of toxic units (TU)). Three concentrations of the metal mixture were tested in replicate systems, and the effects of several parameters were compared with those in the control systems. All systems received a single addition of nutrients at the start of the experiment. Since background levels of the metals were present in the collected water, initial concentrations were somewhat higher than intended. A sum of the toxic units (STU) equal to 1.0 corresponds to a 50% reduction in the population growth of *D. magna* in the laboratory. The range of tested initial STU levels was 0.4 (controls), 0.6 ('low'), 0.8 (median), and 1.3 (high).

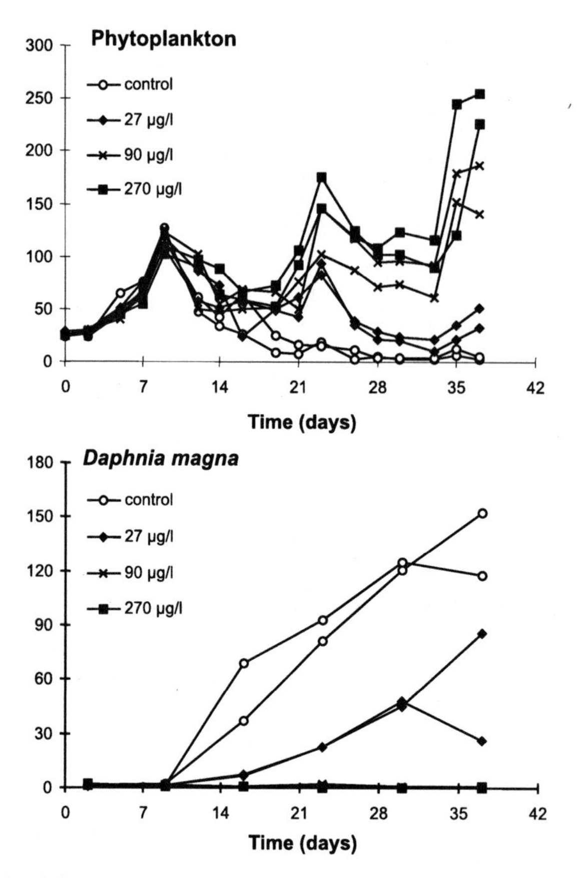

Fig. 3.2. Decreased daphnid development (bottom) due to 3,4-dichloroaniline results in a high algal density (top). Modified from Jak et al. 1998

Despite the fact that the metal concentrations may have caused toxic effects on certain algal species, an overall increase in the chlorophyll-a concentrations was observed. Chlorophyll-a concentrations increased with the added metal concentrations (Fig. 3.3, top). At the beginning of the experiments, edible green algal species (mainly *Scenedesmus sp.*) dominated the algal community. This persisted in the controls and low dosed systems. However, within 4 weeks a shift occurred in the median and high dosed systems towards filamentous cyanobacteria species. These cyanobacteria accounted for the high chlorophyll-a concentrations observed at the end of the experiment. The dose-related increase in the phytoplankton density was clearly induced by toxic effects upon the zooplankton community. While *Daphnia magna* reached high densities in the controls and low dosed systems (Fig. 3.3, bottom), population increase was clearly inhibited by the median metal level. At the highest metal concentration, the population was decimated.

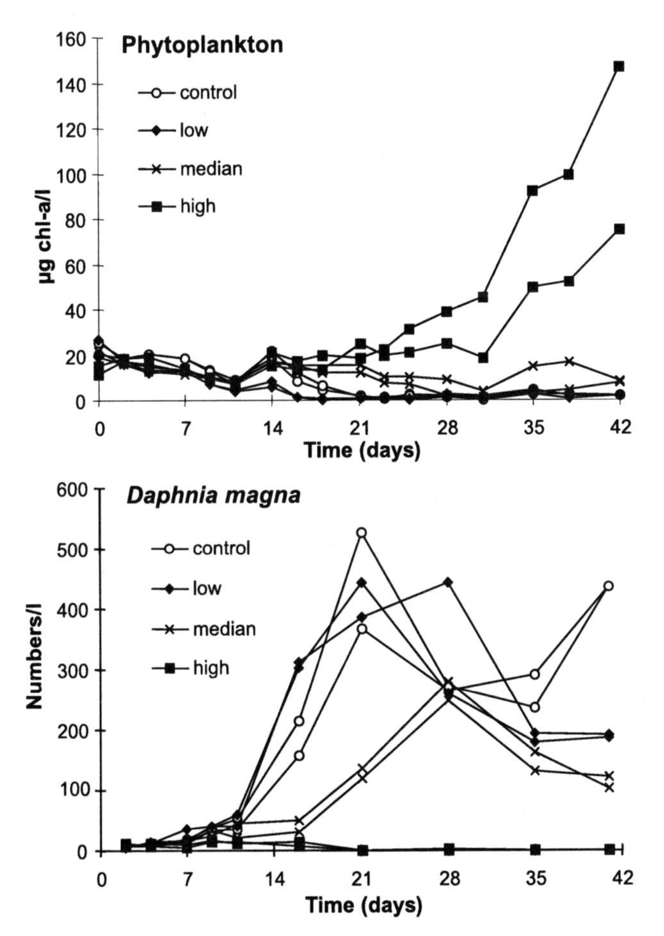

Fig. 3.3. Decreased zooplankton development (bottom) due to a mixture of metals, resulting in a high algal density (top). Modified from Jak et al. 1996

Sensitivity of Different Zooplankton Species

Cladoceran species were clearly the zooplankton group most sensitive to DCA and metals. A comparison of the EC_{50}s , for the population response by different cladoceran species (Fig. 3.4), shows that the smaller species, having lower incipient limiting levels for food, were more vulnerable to toxic substances than the large opportunistic species that dominated systems with high primary production. Differences in relative sensitivity corresponded well with the life history strategies as defined by Romanovsky (see Sect. 2.1). The "violent" standard test species, *D. magna,* was relatively more tolerant than smaller "explerent" species, such as *D. cucullata* and *Ceriodaphnia*, while the "patient" *Bosmina* species was the most sensitive to DCA (this species was absent in the metal experiment). Thus, the sensitivity ranking, also related to the ranking of the species on the basis of the incipi-

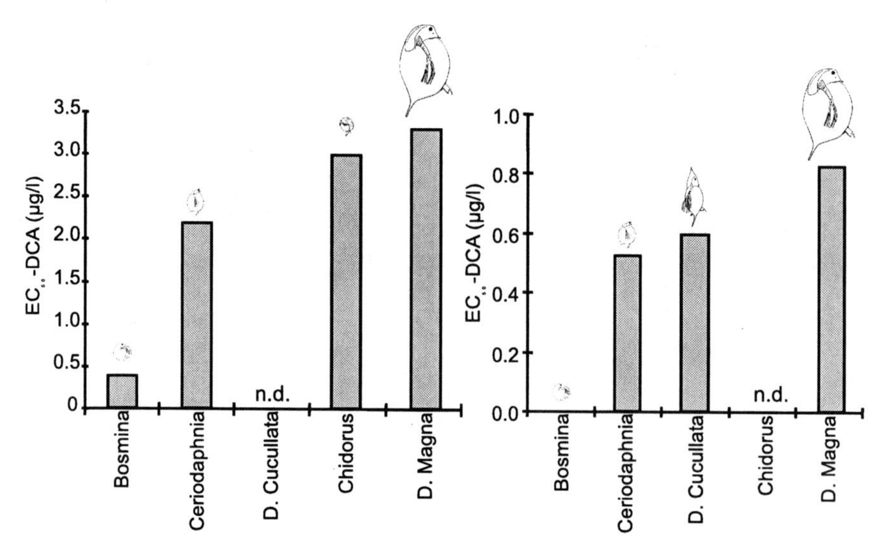

Fig. 3.4. Relative sensitivity of cladoceran species to 3,4-dichloroaniline (DCA), and a metal mixture (data from Jak et al. 1996 and 1998, respectively)

ent limiting level for food (see Sect. 2.1). *Chydorus sphaericus*, a small species that is relatively tolerant towards toxic substances, is an exception. However, this species has a different feeding mode, grasping larger food particles, such as detritus and even cyanobacterial colonies, and is, therefore, not dependent on food levels in the pelagic compartment.

The response of copepods and rotifers to toxicants was a result of both harmful toxic effects (direct) and beneficial effects from reduced competition (indirect) with the more efficiently grazing cladocerans. Figure 3.5 and Fig. 3.6 show, in greater detail, the population development of individual species in both the control conditions and the DCA and metals spiked conditions. For DCA, data of a second experiment (0; 2.7 and 27 µg/l DCA at nutrient enriched and poor conditions) were also included. Copepods appeared to be much less sensitive to DCA and metals than cladocerans, showing higher densities than the controls at low levels of DCA as a result of reduced competition for food. Rotifers were roughly as sensitive as daphnids to metals (with *Keratella quadrata* being the most sensitive). All rotifers showed an opportunistic growth in population due to reduced cladoceran feeding and the subsequently increased algal density by the end of the experiment. In the DCA studies, *K. quadrata* and *K. cochlearis* showed the opposite responses: *K. quadrata* showed a competitive advantage at low DCA levels, and toxic reduction at higher DCA levels, while *K. cochlearis* experienced toxic reduction at low DCA levels and a competitive advantage at high DCA levels.

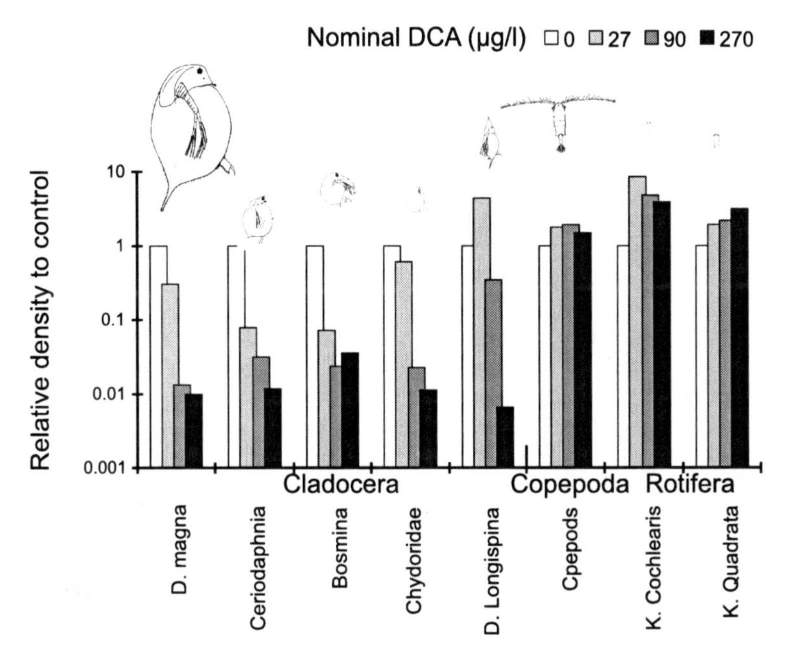

Fig. 3.5. Population response by the major zooplankton species to different levels of DCA, expressed as a fraction of the control population (data compiled from Jak et al. 1998)

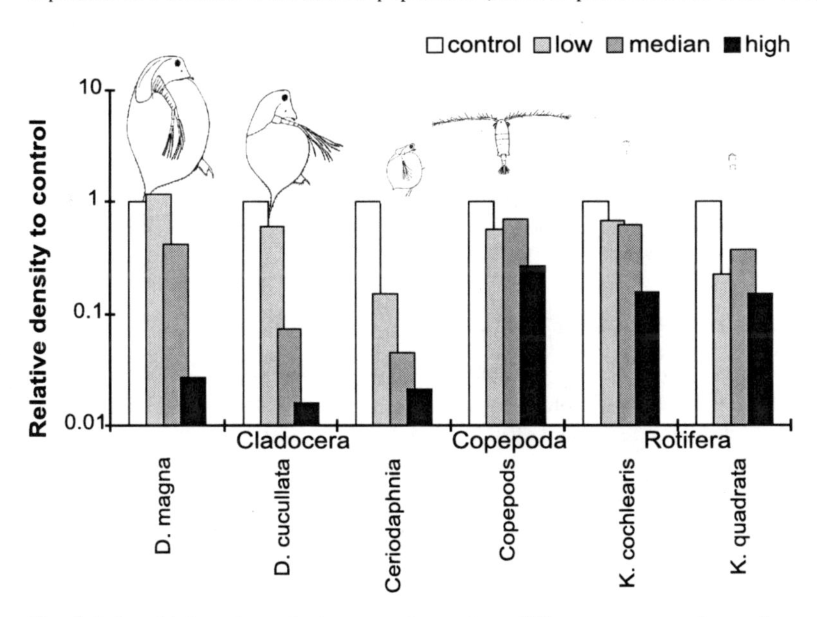

Fig. 3.6. Sensitivity of zooplankton species to three different concentrations of a metal mixture. The population density is expressed as a fraction of the density in the control systems (from Jak et al. 1996)

The opportunistic response of the copepod and rotifer species to reduced competition for food at moderate toxicity levels could not prevent an increase in phytoplankton densities and a shift towards cyanobacteria species. This shift was probably induced by the reduced light conditions due to high algal densities. From these observations it can be concluded that, in a multispecies zooplankton community, the more efficient grazers at low trophic (food) conditions (i.e., small cladoceran species) seem to be more vulnerable to toxic effects than the larger cladoceran *D. magna*, a species adapted to more eutrophied water bodies. Opportunistic copepod and rotifer species seem to benefit from the reduced daphnid grazing.

It was further concluded, from the second DCA experiment, that both the effects of DCA and the diminished control exerted by the zooplankton on the phytoplankton were intensified under nutrient enriched conditions.

This conclusion cannot be explained by intrinsic differences in sensitivity between zooplankton species, and seems to be related to changes in competitive relationships within the zooplankton community. The species that can adapt to the lowest food densities normally has a competitive advantage. When toxic stress affects all species, the reduction of competitive pressure (resulting in a reversed effect on population development) is larger for weak competitors. Therefore, toxic stress will most obviously affect the population development of strong competitors with low food thresholds. As a consequence, the critical algal threshold of the zooplankton community as a whole increases, resulting in higher algal densities and favouring the species adapted to eutrophicated conditions. Odum (1985) formulated the hypothesis that toxicants and other stress factors may induce a shift in the community towards a dominance by opportunistic taxa, resulting in reduced carbon efficiency and energy transfer. A pollution induced shift from conservative zooplankton species to opportunistic ones was, generally, supported by data reviewed by Havens and Hanazato (1993).

The plankton enclosure experiments described above clearly revealed that the addition of nutrients, which provided favourable conditions for algal growth, did not result in increased algal densities, unless toxic effects hampered the development of the dominant grazers (i.e., cladoceran species).

Comparison with a Field Observation

The observations made during the course of the enclosure studies concerning response to metal stress, resemble those made in the field during the restoration of Lago Orta, a volcanic lake in Central Italy. Heavy industrial pollution (viz. emissions of waste water polluted with metals (copper) for several decades) had resulted in the disappearance of almost all forms of life in the eutrophied lake. Metal concentrations in the lake have decreased due to the cleaning-up of emissions and intervention by liming. This has resulted in a change in the plankton community. *Brachionus* was the only zooplankton species that survived under the eutrophied conditions. Over the years, more species have recovered: *Keratella* rotifers recovered first, followed by copepods, then followed by Chydorids and *D. magna* and then followed by "explerent" daphnids like *D. longispina* and *Simocephalus*. The succession was completed with the recovery of *Ceriodaphnia* and *Bosmina*. In the same period, the average summer chlorophyll-a concentrations gradually decreased from more than 50 µg/l to under 5 µg/l (personal communications: Guiliano Bomoni; Univ. Di. Bologna).

Indirect Effects of Pollutants on Benthic Communities

Shifts in the plankton community may result in subsequent shifts in the benthos community. This was observed in a mesocosm experiment (5 m^3, including 30 cm natural sediment and natural water), which was set up to test the effects of the organophosphorous insecticide dimethoate. The community included natural plankton and several introduced macro-invertebrates and water plants (Foekema et al. 1996). Dimethoate was applied in 6 concentrations, ranging from 0–3000 µg/l. Concentrations hardly decreased during the 28 day duration of the experiment. Dimethoate substantially reduced the density of cladocerans in the three highest test concentrations, while the cladoceren:copepod ratio was already affected at 30 µg/l. As a result of decreased zooplankton grazing, increased algal densities were observed for the three highest test concentrations. The most sensitive response measured, however, was the enhanced growth of the fingernail clam (*Sphaerium corneum*), a filter feeding bivalve, which apparently benefitted from improved food conditions. Enhanced growth of the filter feeding bivalve *Sphaerium corneum* was already significant at concentrations as low as 10 µg dimethoate per litre (Fig. 3.7).

Effects of Polluted Sediment on Plankton Dynamics

The previous experiments showed the effects of selected pesticides and metals on zooplankton grazing. Similar effects were observed in the semi-natural situation of natural sediment, from a polluted and a relatively pristine location, which was left for several years in outdoor mesocosms (2 m^3, 30 cm sediment layer) (Foekema et al. 1998). The polluted (Lake Ketelmeer) and unpolluted (Lake Oostvaardersplassen) sediments were very similar with regard to their physical and chemical (e.g. nutrient) characteristics, but differed in their content of pollutants (esp. several heavy metals and PAHs).

Special attention was paid to the plankton dynamics in late winter and early spring, and sampling was started after the ice cover had melted. The most noticeable effects included the higher phytoplankton levels in spring in the water column above the polluted sediment, and the delayed development of cladocerans Fig. 3.8). The species composition of the zooplankton also differed. In the unpolluted system, *Daphnia longispina* dominated while in the polluted system the small *Bosmina longirostris* predominated. Copepods, which were most numerous after the winter, were not affected.

The study demonstrates that a delayed development of cladocerans in spring may result in enhanced phytoplankton levels. This delay may induce an aquatic ecosystem to shift from a clear water state after the spring bloom into a eutrophied state in which high phytoplankton levels persist throughout the summer period.

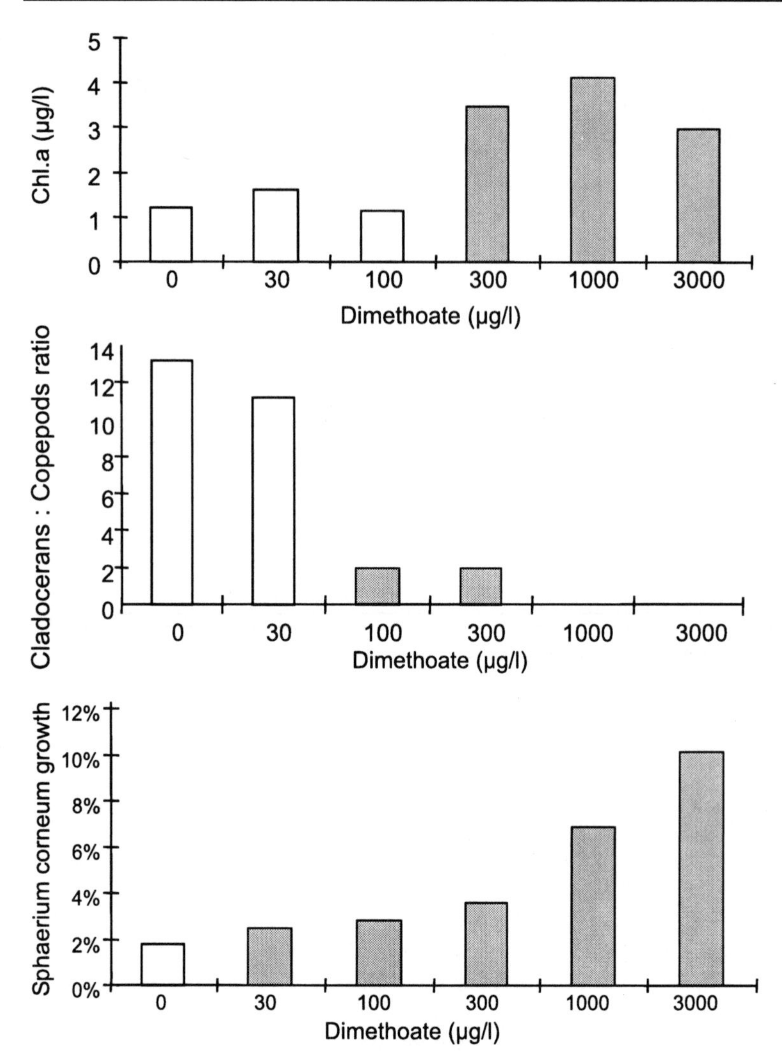

Fig. 3.7. Response of phytoplankton and two groups of phytoplankton herbivores (zooplankton and mussels) exposed to a concentration series of dimethoate. The dark bars differ significantly (α= 0.05) from the controls (0 µg/l). Experiment carried out in 5 m^3 outdoor ponds systems (Foekema et al. 1996)

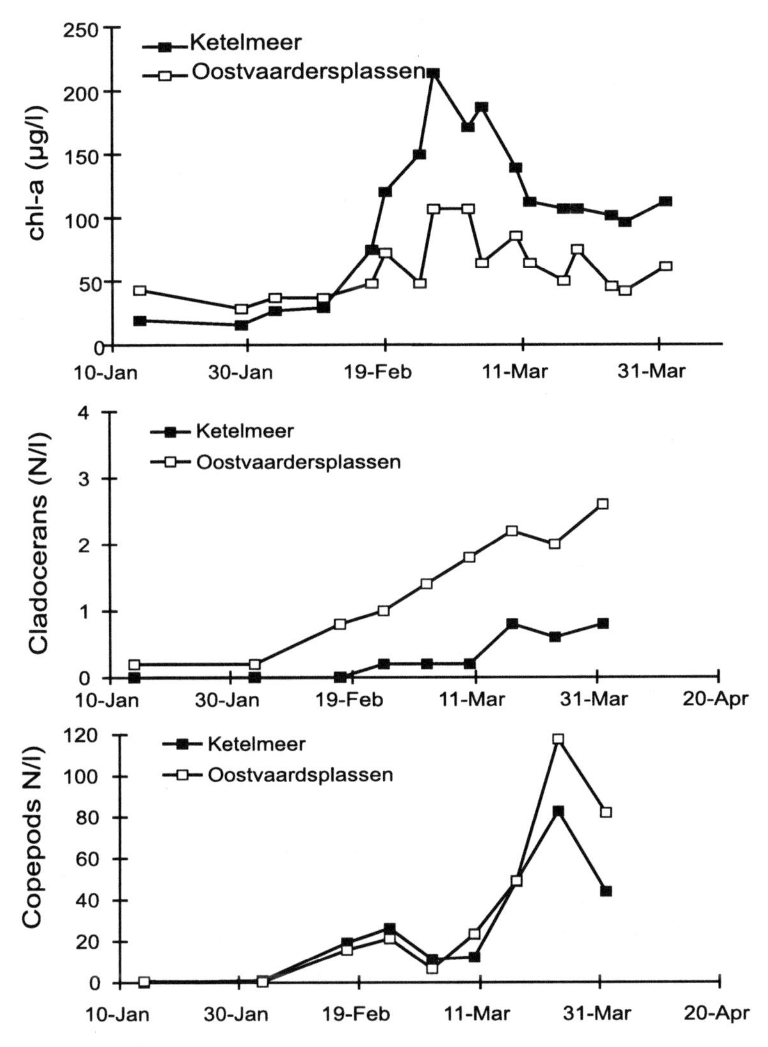

Fig. 3.8. Development of the algal density , cladoceran density and copepod density in mesocosms with polluted sediment ("Ketelmeer") compared to unpolluted sediment ("Oostvaardersplassen") (Foekema et al. 1998)

3.3 Daphnid Grazing Effectiveness in Response to Toxicant Exposure

Plankton Eco-assay as Ecotoxicity Test

In the previous section, we have seen that toxic substances may affect the zooplankton community in such a way that its ability to control the algal density by grazing is reduced, resulting in prolific algal development and, in some cases, in other shifts in the ecosystem (e.g., stimulation of benthic filterfeeders). In this section, the effect of toxicants on daphnid grazing will be studied in more detail, under controlled, laboratory conditions, using the plankton eco-assay (Sect. 2.2).

A sensitive and reliable ecotoxicological test on grazing effectiveness requires optimal grazing conditions: i.e. an initial daphnid density that is in equilibrium with the initial algal density in order to allow grazing control of the algal production (see Sect. 2.2). When the daphnid density is too high in comparison to the algal production level, food is limiting and grazing is determined by the food concnetration. When the daphnid density is too low in comparison to the algal production level, grazing effectiveness cannot be determined due to uncontrolled exponential algal biomass development (Fig. 3.9). The test must be rejected when the data from the control systems (no toxicants) indicate that algal production and grazing are unbalanced (algal depletion or exponential algal development).

In practice, the method development for the plankton eco-assay test required a build-up of experience in order to achieve balanced conditions between algal production and grazing in terms of initial algal density for specific algal species or assemblages and initial daphnid density and species composition (see also Chap. 2.3). We executed more than 30 plankton eco-assays during the method development period. The results obtained from 10 typical eco-assays (Table 3.3) will be used for the study of toxic effects on daphnid grazing in more detail.

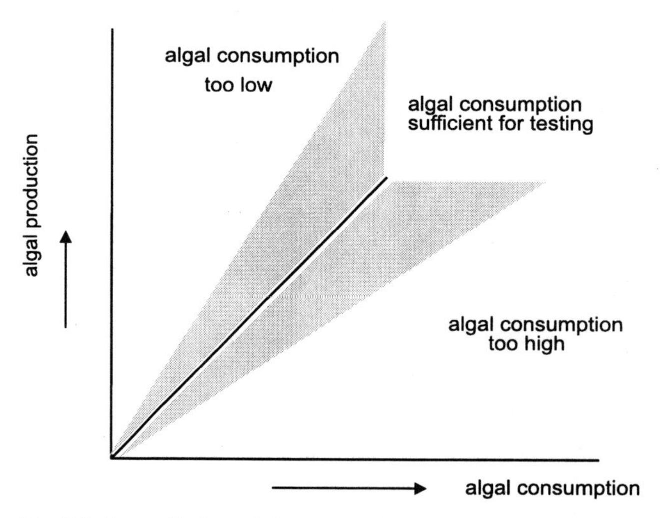

Fig. 3.9. A good balance between algal production and algal consumption by daphnids is a requirement for an informative daphnid grazing test

Table 3.3. Lowest observed effect concentrations (LOECs) of various substances for algae production and daphnid grazing as determined in the plankton ecoassay. The LOEC-algae refers to the lowest test concentration that caused a significant ($p<0.05$) reduction of the algal growth rate in systems without daphnids. The LOEC-daphnids refers to the lowest test concentration that caused a significant ($p<0.05$) elevation of the algal growth rate in systems with daphnids. 'n.t' (not tested) indicates that in this test no systems 'without daphnids' were treated with the test substance

Test#	Toxicant	Test medium	Algal species	Daphnid species (n per litre)				LOEC algal production	LOEC daphnid grazing
				Dm	Dl	Sv	Cd	µg/l	µg/l
1	Dimethoate	Natural	Natural	2	2	2		n.t.	18
2	Dimethoate	DSW	Chlorella	1		2	4	n.t.	100
3	Dimethoate	DSW	Chlorella	3		3		n.t.	320
4	Dimethoate	OECD	Chlorella/ Raphidocelis	5	5	5		>320	320
5	Fluoranthene	OECD	Raphidocelis	5	5	2		>56	6
6	Fluoranthene	OECD	Chlorella	10				32	10
7	Cadmium	OECD	Chlorella	2			5	>100	100
8	Copper	OECD	Chlorella	4			5	> 30	30
9	Copper	OECD	Raphidocelis	5	5	5		32	> 32
10	NaCl	Natural	Natural	2		2	3	n.t.	3 200 000

*Note: Dm=*Daphnia magna*; Dl=*Daphnia longispina*; Sv=*Simocephalus vetulus*; Cd=*Ceriodaphnia dubia*

Reproducibility

The observed effect concentrations (LOEC) from 3 of the 4 tests with the insecticide dimethoate were in the same range (100–320 µg/l). A lower LOEC of 18 µg/l was found in one test with natural water and phytoplankton. This was most likely due to variation in the test conditions. As mentioned above, the balance between algal production and grazing is very subtle and it had a strong impact on the sensitivity of the eco-assay. The algae-daphnia community was apparently in a sensitive balance when the lowest LOEC was determined. Over the course of the years during which the the method was developed, it became clear that it is very difficult to produce this subtle/sensitive balance in a short-term experimental set-up. In the field situation however, this balance is most likely to be maintained continuously since all species are competing for the scarce recourses and are thus continously 'living on the edge'.

Example: Dimethoate (Test#2)

The results of this test are shown in Fig. 3.10 and Fig. 3.11. The presence of a dimethoate concentration above 32 µg/l did affect the grazing efficiency of the daphnids, which resulted in substantially higher nett algal growth rates. At dimethoate concentrations of 10 and 32 µg/l the algal development was controlled by daphnid grazing and was comparable to the control situation (0 µg/l dimethoate).

Simocephalus vetulus was the most dominant species in the control systems at the conclusion of the experiment. However, at a dimethoate concentration of 10 µg/l, the density of *S. vetulus* started to decline and the obviously less sensitive species *D. magna* was able to increase in density. Hardly any *Simocephalus* were observed in the samples at a concentration of 100 µg/l, while *D. magna* still seemed unaffected. *Ceriodaphnia dubia* was only present in small, insignificant numbers.

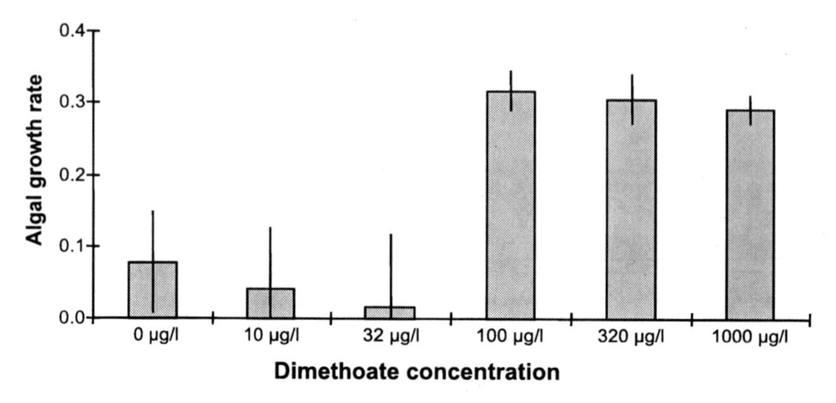

Fig. 3.10. Relative growth rate of algae in plankton eco-assays with daphnids in a concentration series containing the insecticide dimethoate (test#2). The test medium was Dutch Standard Water, enriched with 0.5 mg/l PO_4-P and 3.6 mg/l NO_3-N. Initial algal density was 20 µg/l chl-a

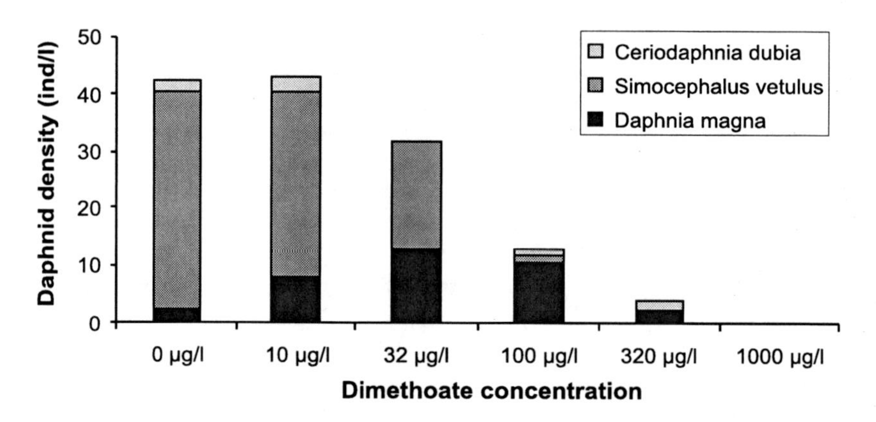

Fig. 3.11. Density of daphnids at the end of the plankton eco-assay in a concentration series containing the insecticide dimethoate (test#2). Initial algal density was 1 *D. magna*, 2 *S. vetulus* and 4 *C. dubia* per litre

The daphnid community lost control over the algal development at a dimethoate concentration of 100 µg/l. The most sensitive species *Simocephalus vetulus*, seemed to have played a key role in the control of the algal development in this experiment.

In test#3, *Simocephalus vetulus* was also the more sensitive species, but was less dominant and less crucial, therefore (Fig. 3.12). This might explain the higher LOEC for the toxic reduction of the grazing effectiveness: 320 µg/l in comparison to 100 µg/l in test#2.

In test#1 with natural phytoplankton, no significant changes in daphnid development were observed at dimethoate concentrations of up to 56 µg/l (Fig. 3.13), although reduced grazing effectiveness was observed at 18 µg/l dimethoate (LOEC) and higher. This indicated a sublethal grazing reduction or "toxic anorexia" of the daphnids, which had not yet resulted in reduced reproduction.

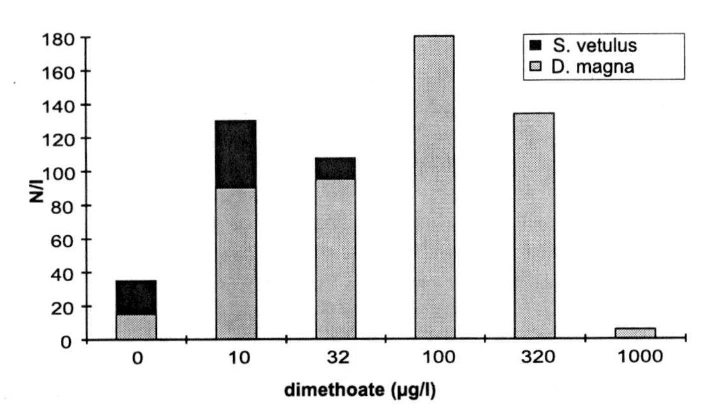

Fig. 3.12. Density of daphnids at the end of the plankton eco-assay in a concentration series withf the insecticide dimethoate (test#3)

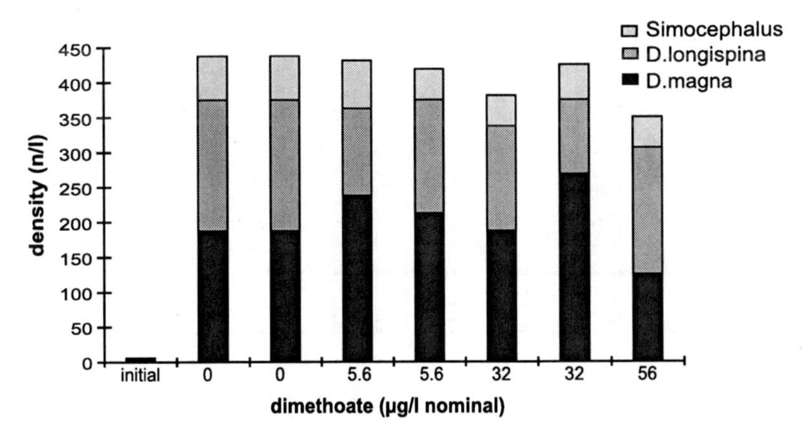

Fig. 3.13. Daphnid density at the end of the plankton eco-assay in a concentration series with the insecticide dimethoate (test#1)

The LOECs recorded in the plankton eco-assay for grazing reduction were in the range of the NOEC, recorded in a 16 day reproduction test with *Daphnia magna,* of 29 µg/l dimethoate (Ordelmans et al. 1994).

Example: Fluoranthene (Test#5)

Dimethoate is an insecticide known to be toxic to zooplankton but is insignificantly so to algae. A test with the PAH fluoranthene is presented here as an example of the results of a plankton eco-assay carried out with a less specific toxicant (Fig. 3.14, Fig. 3.15).

The algal growth rates in the containers without daphnids revealed the effect of fluoranthene on the algae. A significant decrease in the growth rate was observed at 56 µg/l fluoranthene.

The daphnids were capable of substantially reducing the algal growth rate in the control situation without fluoranthene (Fig. 3.14). However, even at the lowest test concentration of 5.6 µg/l this control was completely lost and the algal growth rate was comparable to that in the containers without daphnids. The algal growth rate was further reduced at fluoranthene concentrations of 32 and 56 µg/l. However, this was the result of the direct impact of the fluoranthene on algae and not due to daphnid grazing, since the daphnids were unable to survive at these test concentrations.

It is remarkable that a reduction of less than 25% of the daphnid population at the lowest test concentration caused the complete loss of the algal development in the control . All three daphnid species seemed to be equally sensitive to fluoranthene. A further reduction in the daphnid density at higher test concentrations was not reflected in a higher algal growth rate. This emphasises the subtlety of the balance between algal production and daphnid grazing.

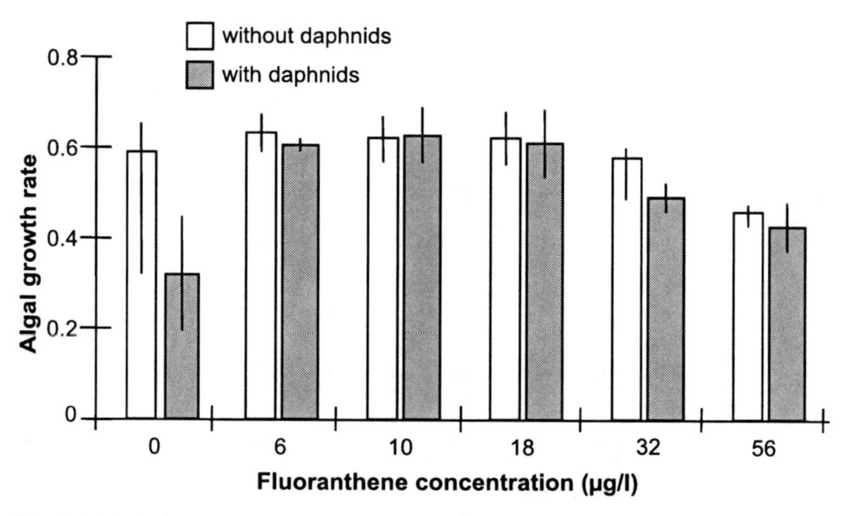

Fig. 3.14. Relative growth rate of algae in a plankton eco-assay with or without daphnids in a concentration series of the PAH fluoranthene (test#5). OECD test medium; initial algal density 10 µg/l chl-a. The algae used was *Raphidocelis subspicata*

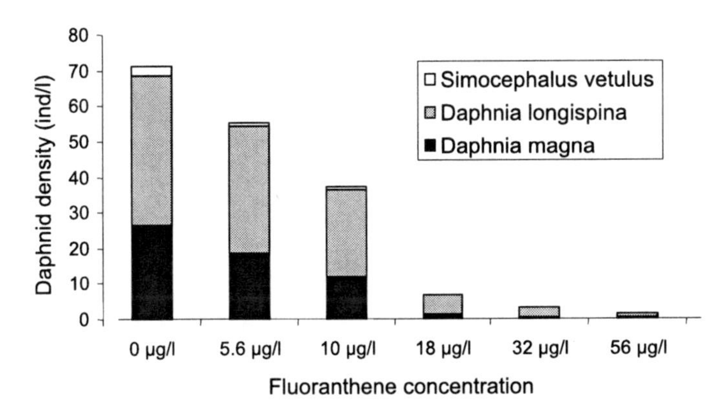

Fig. 3.15. Daphnid density at the end of the plankton eco-assay in a concentration series with the PAH fluoranthene (test#5). Initial density: 5 *D. magna*, 5 *D. longispina* and 1.5 *S. vetulus* per litre

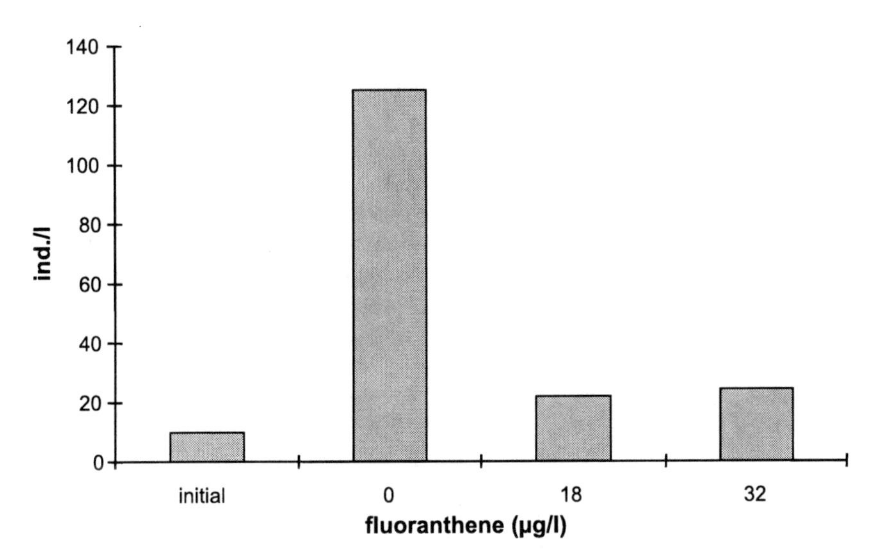

Fig. 3.16. Daphnid density at the end of the plankton eco-assay in a concentration series with the PAH fluoranthene (test#6)

In test#6, the other test conducted with fluoranthene, the reduced daphnid grazing correlated with reduced *D.magna* development.

The observed LOECs recorded in the plankton eco-assay for grazing reduction is low compared to reported chronic NOECs for daphnid reproduction of approx. 100 µg/l fluoranthene (Suedel and Rhodes 1993; Suedel et al. 1996). However, Taylor (pers. comm.) observed inhibited grazing by daphnids from video observations at concentrations below 10 µg/l. It should be noted that toxicity of fluoranthene can be enhanced by illumination (phototoxicity).

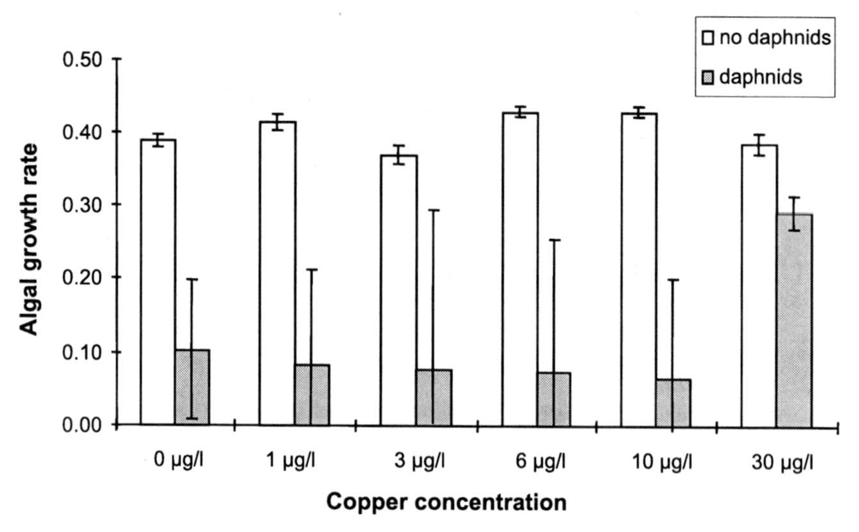

Fig. 3.17. Relative growth rate of algae in a plankton eco-assay with or without daphnids in a concentration series with copper (test#8)

Example: Copper (Test#8)

The copper test was executed under conditions similar to the fluoranthene test#5. Copper is known to be toxic to algae, but no effect on *Raphidocelis* growth rate was observed within the test range (1–30 µg/l, see Fig. 3.17). Daphnids were able to substantially reduce the growth rate by grazing, except in the highest oncentration (30 µg/l), where daphnid grazing was reduced.

Despite the recorded reduction in algal grazing, the daphnid development was not affected at 30 µg/l copper (Fig. 3.18). Again, this was a clear indication for "toxic anorexia" at relatively low effect concentrations, although comparible to an average chronic NOEC for daphnid reproduction of 8.5 µg/l (De Bruijn et al. 1999).

Example: Potassium Dichromate

Potassium dichromate ($K_2Cr_2O_7$) is the standard reference toxicant in aquatic ecotoxicity tests. However, a complex response was recorded in the plankton-eco-assay Fig. 3.19). Reduced algal growth due to $K_2Cr_2O_7$ toxicity was observed over the first four days. Thereafter, toxicity induced grazing reduction by daphnids resulted in an opposite response: an increase in algal density with $K_2Cr_2O_7$ concentration. Despite clear dose-effect related inhibition of algal growth, at the end of the test, an increase in algal density with toxicant concentration was observed as a result of reduced grazing. This example illustrates that toxic effects on algal growth that might be apparent in acute algal toxicity tests, might be overshadowed by a toxic reduction of daphnid grazing over the long term.

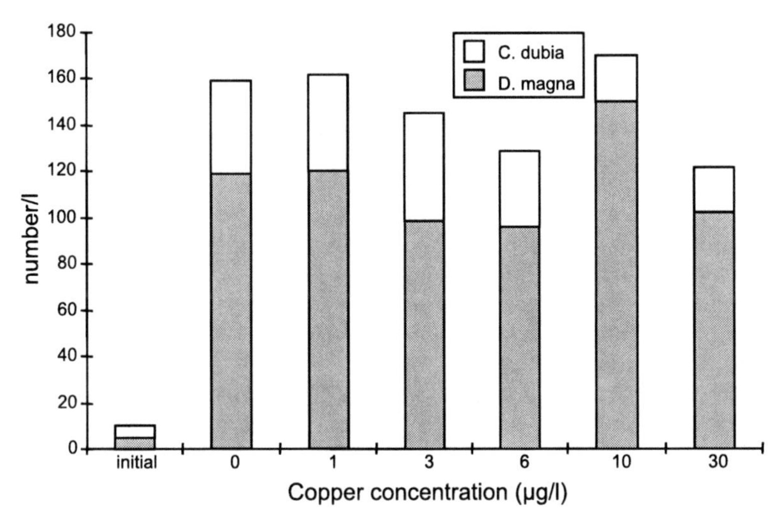

Fig. 3.18. Daphnids density at the end of the plankton eco-assay in a concentration series with copper (test#8)

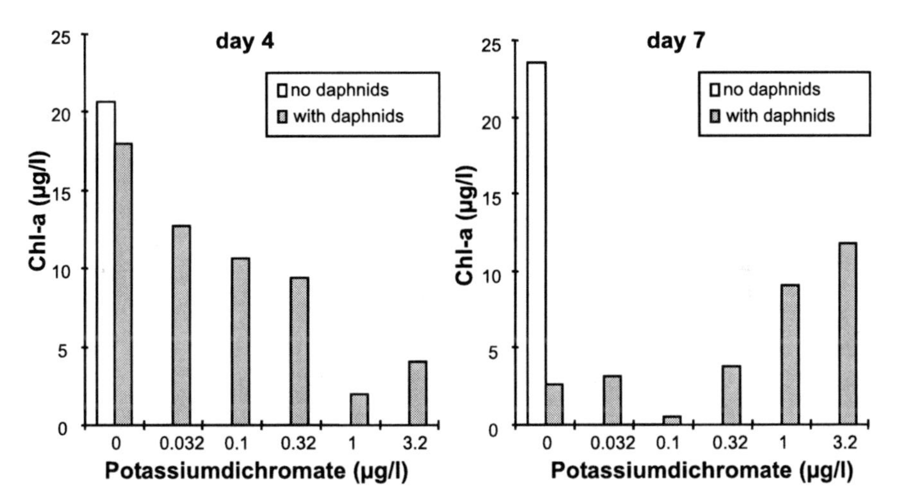

Fig. 3.19. Algal densities over a concentration series with potassiumdichromate, determined in an eco-assay conducted with natural surface water (Lake Geestmerambacht) with a flagellate and diatom dominated algal community (initial chlorophyll-a concentration of 23 µg/l). Initial daphnid densities were 3 individuals of *D. magna* and 2 individuals of *S. vetulus* per litre. The algal densities (measured as chlorophyll-a) on day 4 and day 7 are presented. Algal development without daphnids was only tested in the references

Synopsis

Daphnid grazing effectiveness is a sensitive ecotoxicological endpoint, yielding effect concentrations in a 7 day test comparable to those observed in 21–28 day chronic reproduction tests. Even lower effect concentrations are found in some cases.

The toxic effects were observed at the lowest effect concentrations irrespective of daphnid density development, indicating a reduction in the feeding of the initial daphnid population. This toxic effect is commonly referred to as "toxic anorexia".

Over the longer term, "toxic anorexia" will result in reduced reproduction and subsequent delayed population development, as observed in chronic reproduction tests and plankton enclosure studies.

The fact that the observed effect concentrations for daphnid grazing effectiveness were comparable to the lowest observed effect concentrations for the most sensitive (often indirect) responses in outdoor mesocosm studies confirmed the theory that daphnid grazing is a critical ecological process in the matter of the response of aquatic ecosystems to toxicants.

4 Field Observations of Daphnid Grazing

4.1 Two Different Lakes in Holland

From mesocosm studies and plankton eco-assays examining toxicant exposure, it has become clear that the grazing effectiveness of daphnids is an important factor in plankton dynamics, and that the grazing effectiveness can be reduced by toxicant loading. In this chapter, the relevance of daphnid grazing in the field situation will be demonstrated on the basis of field surveys carried out in two Dutch lakes: Lake Geestmerambacht and Lake Amstelmeer.

Lake Geestmerambacht and Lake Amstelmeer are two moderately deep, man-made lakes in the province of North-Holland in the Netherlands (see Fig. 4.1). Both lakes were studied quite intensively during the nineteen nineties, with particular attention paid to their plankton communities. Comparative eco-assay studies were performed with water from both lakes in order to acquire an improved understanding of the variation in the grazing effectiveness of daphnids, and biotic and abiotic factors that may influence it .

Morphology

Lake Geestmerambacht was created from 1967 to 1979 as a consequence of sand excavation. The lake has a surface area of 70 ha and an average depth of 11 metres, with a deep area of 20–21 metres in the center . It is an occasional reservoir (8 Mm^3) for excessive polder water. The water residence time is more than 15 years. The surrounding area is used for recreation and pasture for cattle farming. The water is mildly brackish (salinity 0.25‰). The lake is monomictic, with stratification occurring in summer during the period from May–June to September–October (WL 1996; Van Dokkum and Van der Veen, 2000).

Lake Amstelmeer is a former tidal channel (Amsteldiep) of the Wadden Sea tidal area. In 1925, a dam was constructed separating the channel from the sea, thereby creating the lake. It has a surface area of 650 ha and is moderately deep with an average depth of 4.5 metres and a central section that is 10–16 metres deep. Lake Amstelmeer is an operational reservoir (29 Mm^3) for superfluous polder water from a catchment area consisting of 24 000 ha of polders in agricultural use (flower bulb cultivation, arable land). The residence time of the water is 2–3 months. The lake has slightly brackish water (salinity: 0.5–1.5‰).

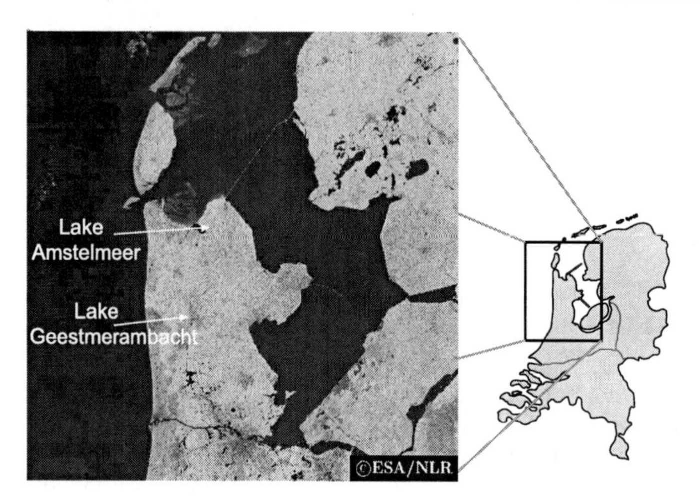

Fig. 4.1. Location of Lake Geestmerambacht and Lake Amstelmeer in the Netherlands

Water Quality

Both lakes are eutrophic: the average total P concentration is 0.45 mg/l in both lakes. The Kjeldal N concentration is higher in Lake Amstelmeer (2.3 mg/l) than in Lake Geestmerambacht (1.4 mg/l).

Lake Geestmerambacht does not have a permanent eutrophied character, and turbidity is generally low with a secchi-depth of 50–320 cm; and chlorophyll-a concentrations ranging from < 8 µg/l during the clear water phase which follows a short spring bloom of up to 185 µg/l during the cyanobacteria blooms that are regularly observed during August–September, (Van Dokkum et al. 1999; Van Dokkum and Hoornsman, 2000; Foekema and Van Dokkum, 2000; Holthaus et al. 2001).

In contrast, Lake Amstelmeer has a permanent eutrophied character (*senso lato,* see Chap. 1): a high turbidity, secchi-depth of 30–140 cm; no submerged vegetation; and a chlorophyll-a concentration typically ranging from 50 up to more than 200 µg/l (Fig. 4.2). A clear water phase is not reached.

4.2 The Plankton Dynamics in Lake Geestmerambacht

Phytoplankton

The phytoplankton dynamics in Lake Geestmerambacht show a typical seasonal pattern (see Fig. 4.3).

In winter, the chlorophyll-a concentration is low and the transparency of the lake is high. In the spring, when the water temperature and day-length increase, phytoplankton starts to develop and the spring bloom can reach chlorophyll concentrations up to 185 µg/l. Diatoms and green algae dominate the plankton (AquaSense 1996) (see Fig, 4.4). After the spring bloom has collapsed, a clear water

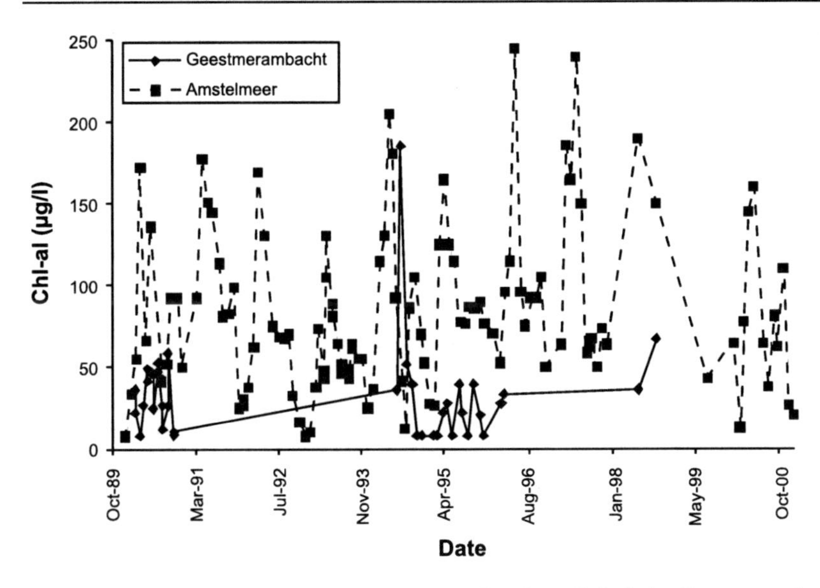

Fig. 4.2. Algal density (expressed as chlorophyll-a, in µg/l) in Lake Geestmerambacht and Lake Amstelmeer in 1990–2000

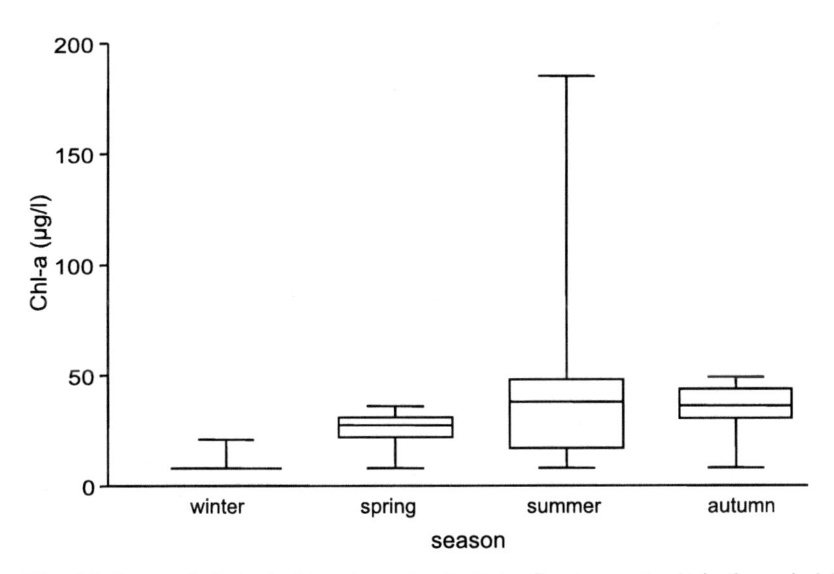

Fig. 4.3. Seasonal phytoplankton dynamics for Lake Geestmerambacht in the period 1990–2000 (box-whisker plots). See Fig. 4.10 for comparison

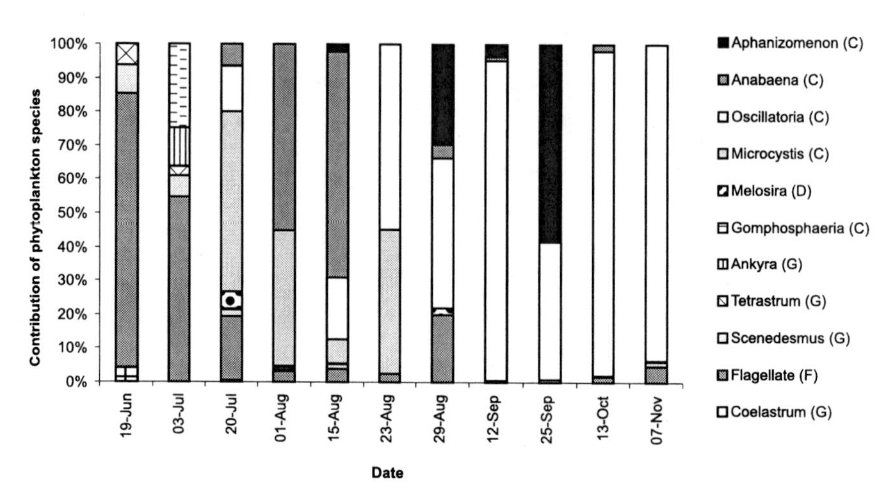

Fig. 4.4. Phytoplankton composition in Lake Geestmerambacht water in 2000. Abbreviations in the legend: (G)=green algae; (D)=Diatom; (C)=cyanobacteria; (F)= Flagellate

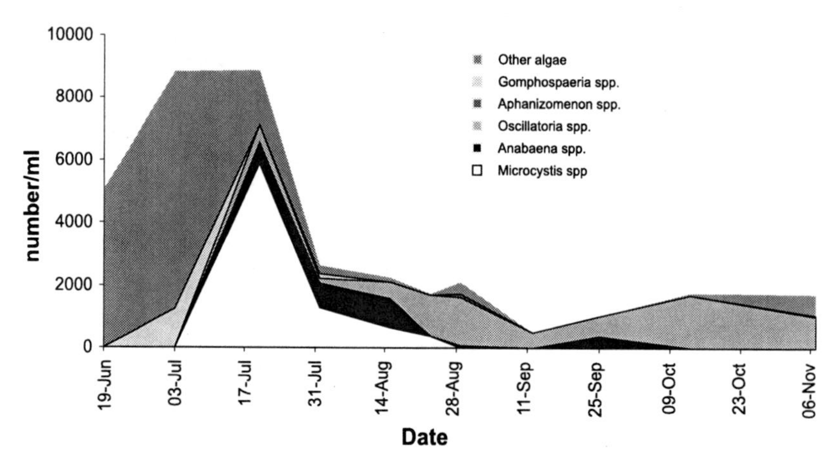

Fig. 4.5. The summer/autumn bluegreen algae bloom in Lake Geestmerambacht in 2000. In July, bluegreen algae (esp. Microcystis) began to dominate the plankton. In mid August, floating mats of Microcystis aeruginosa were observed. After this, an Oscillatoria bloom developed. Other algae: mainly flagellates and green algae. Data from Holthaus et al. 2001

phase with chlorophyll-a densities < 50 µg/l is reached. During the summer months, however, bluegreen algae begin to develop. The summer bluegreen algae bloom was monitored from 1998 to 2000 (Van Dokkum et al. 1999; Van Dokkum and Hoornsman, 2000; Foekema and Van Dokkum 2000; Holthaus et al. 2001). Bluegreen algae start to develop in June, and at the end of July/ early August the plankton is dominated by cyanobacteria (> 90%). The bloom continues to the end of October.

The morphology of the lake is probably a major reason for the dominance of cyanobacteria in the late summer and autumn. Lake Geestmerambacht is a relatively deep lake, with a depth of 20–22 meters. The lake is monomictic and becomes stratified from ca. June to October. The thermocline is located at a depth of ca. 10 meters in June, and ca. 15 meters in October (AquaSense 1996). During the period of stratification, phosphorus and nitrogen are incorporated in algal biomass in the epilimnion. Due to the death and subsequent sedimentation of the algae, the epilimnion may be depleted of nutrients. In these circumstances, bluegreen algae with the ability to control their own buoyancy have a competitive advantage over algae that cannot do so, because they can find nutrients near the thermocline at night and in the light near the water surface during the day (Chorus and Bartram 1999). *Microcystis*, the dominant cyanobacteria occurring during the summer bloom, is typical for stratified monomictic lakes (Chorus and Bartram 1999).

Zooplankton

The zooplankton in the lake is not studied on a regular basis, and therefore little information is available., An inventory of the zooplankton dynamics was made in 1994 (AquaSense 1996, see Fig. 4.6). Substantial numbers of rotifera and copepods were found from March to November. Cladocerans were present in June and August–November,. The species found in June was *Bosmina coregoni*.The cladoceran community was more diverse in the autumn with *Daphnia hyalina*, *Diaphanosoma brachyurum* and *Bosmina longirostris* also present.

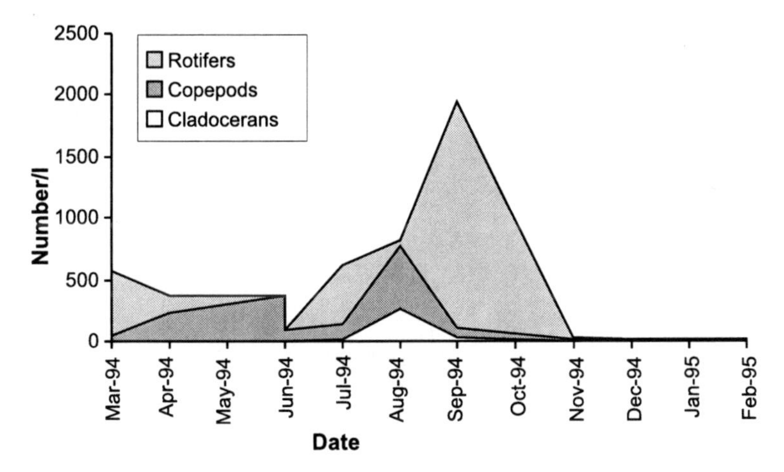

Fig. 4.6. Zooplankton dynamics at one location in Lake Geestmerambacht in 1994. Data from AquaSense 1996

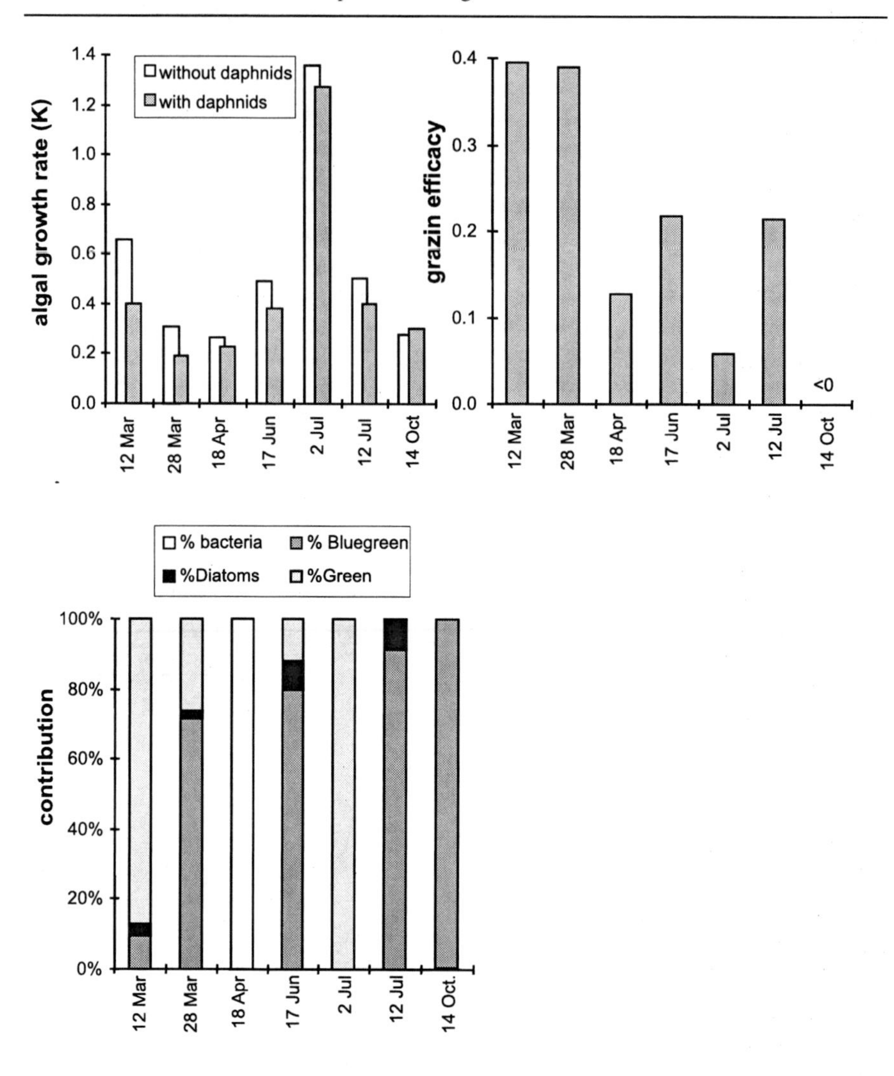

Fig. 4.7. Upper left panel: Development rates of natural algal communities from Lake Geestmerambacht in the plankton-eco-assay, sampled throughout the year. Right panel: The daphnid grazing effectiveness determined in these plankton eco-assay tests at initial daphnid densities of 8 per litre. Lower left panel: The phytoplankton composition

Zooplankton Grazing

In 1996, a series of plankton ecoassays were performed with water from Lake Geestmerambacht, in order to characterise the grazing effectiveness of daphnids in response to the seasonal change in algal composition. The grazing effectiveness (Fig. 4.7) showed a clear seasonal variation with a general reduction in the grazing effectiveness throughout the year. The highest grazing effectiveness (40%) was reached in March. In the summer period the grazing effectiveness fell to below 20%. In the test performed on July 2nd, a grazing effect capable of inhibiting the development of algal density was, apparently, completely absent. The failure of grazing was most likely caused by the extremely rapid growth of the green algae *Chlorococcus.* Moreover, *Chlorococcus* can appear in a broad size range from 10–100 μm, and it is possible that selective grazing on the edible smaller cells could cause a shift of the population towards larger, inedible cells.

This series of tests clearly revealed that the capability of daphnids to control algal growth may vary considerably throughout the year, due to the changes in algal composition. Daphnid grazing throughout the growing season may cause a shift in the algal community to less edible species partly as a result of grazing. However, it also revealed that even blue-green algae could be grazed reasonably effectively (GR% = 10–20%) in these studies with relatively low initial algal concentrations (i.e. *Anabaena sp.* dominating in June).

Daphnid grazing, however, was completely absent during the final test of this series in October. In this test, the algal community consisted entirely of a population of the blue-green *Oscillatoria,* which often exists in filamentous structures which are not easily ingested by daphnids.

It was observed in an experiment using *Oscillatoria* (see Sect. 2.2), that the grazing effectiveness on filamentous cyanobacteria is heavily dependent on the initial daphnid densities.

Daphnid grazing on *Anabaena sp.* in July was also observed as being only marginal (GR% 2–5) at initial densities of 4 or 6 *D. magna* per litre whereas, at initial densities of 8–12 individuals per litre, a grazing effectiveness of 20, to over 50%, was obtained, resulting in moderate control of the *Anabaena* population development.

Resilience of the Plankton Community to Eutrophication

From a lake management point-of-view, it is important to have information on the response of the plankton community to an increase in the nutrient load. The "resilience" of the plankton in Lake Geestmerambacht was studied in a series of three experiments, where nutrients were added to the natural water (including phyto- and zooplankton species and densities) and the response of the plankton was recorded. The results are shown in Fig. 4.8. The algal density increased when increasing amounts of nutrients were added to the water, but the zooplankton was still able to exert a certain amount of top-down control. The chlorophyll-a density was a constant < 80 μg/l.

Another experiment was performed by adding daphnids to the water from which the resident zooplankton community was excluded by sieving. The effect of zooplankton on algal density was demonstrated by carrying out experiments with and without (added) zooplankton. In the systems without daphnids, higher algal concentrations were reached at higher nutrient loads. In the systems with daphnids, however, the algal densities were controlled at much lower levels (Fig. 4.9). The results of the eco-assay strongly resemble the semi-field observations of the chlorophyll-a to P response under daphnid rich and poor conditions. Sarnelle (1992) calculated a chlorophyll to P ratio of 0.5 when few daphnids were present,

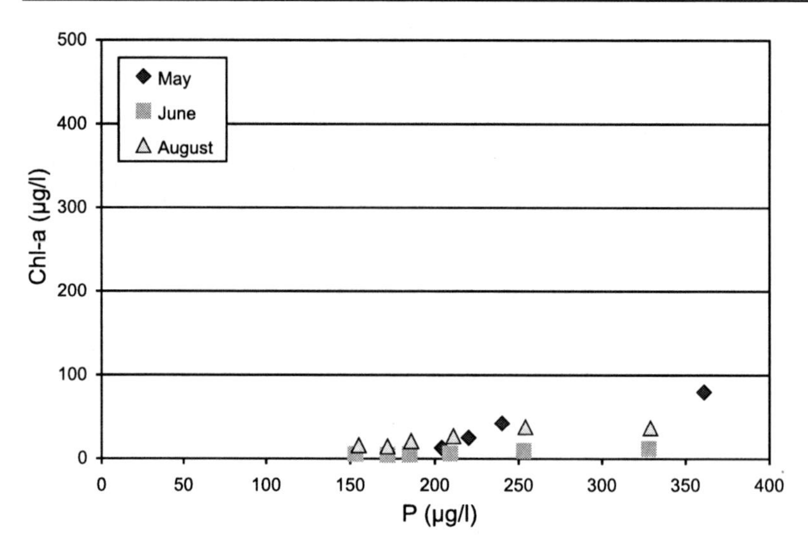

Fig. 4.8. Response of the phytoplankton (expressed as maximum chlorophyll-a concentration measured) after addition of nutrients to natural water from Lake Geestmerambacht in indoor microcosms. The experiments were carried out in 1995. Initial zooplankton densities (cladocerans – copepods) were 17–13 in the May experiment; 45–12 in the June experiment, and 5–3 in the August experiment. The nutrient concentration on the x-axis was natural P + P addition. See Fig. 4.12 for comparison

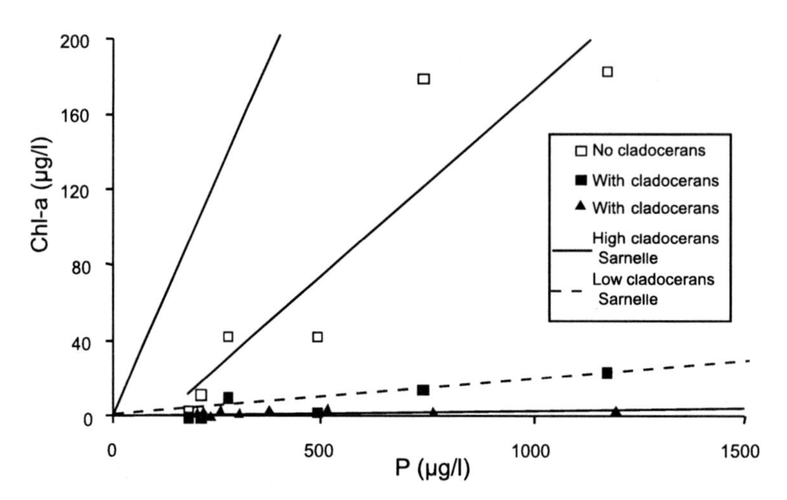

Fig. 4.9. Response of algal density (chlorophyll-a) to increased P-load in systems with and without daphnids after 9–12 days. The lines are chlorophyll to P response under Daphnid rich and Daphnid poor conditions predicted by Sarnell (1992); see text

and 0.02 when daphnid densities were high. In the eco-assay, chlorophyll to P ratio of 0.2 was recorded in the absence of daphnids, although a ratio of 0.02 was also recorded with daphnids present.

Synopsis

In Lake Geestmerambacht, a spring bloom of diatoms and green algae is usually followed by a short clear water phase, in June. During this time, cladocerans, copepods and rotifers, which are able to control algal densities at a low level, are present. Grazing is efficient during this period. Experiments have shown that the zooplankton community can control the phytoplankton production during this period, even when nutrients are added to the system. The fact that a summer/autumn bluegreen algae bloom occurs each year is probably not the result of reduced top-down control. It is rather a result of the stratification of the lake and the subsequent depletion of the nutrient pool in the epilimnion, thereby providing favourable conditions for Microcystis and other bluegreen algae. Cladocerans are present during this bloom, and have been shown to graze on some cyanobacteria (Oscillatoria). However, the zooplankton cannot prevent cyanobacteria from blooming. In late autumn, when the stratification breaks and temperatures drop, the cyanobacteria bloom collapses and a clear winter phase starts.

4.3 The Plankton Dynamics in Lake Amstelmeer

Phytoplankton

The annual dynamics of phytoplankton biomass is characterised by relatively high densities, even in winter (Fig. 4.10).

The phytoplankton is dominated by cyanobacteria, sometimes associated with green algae. Diatoms are only present in low densities during the spring bloom. Dominant cyanobacteria appear in the following order: *Oscillatoria spec.* (autumn/winter), *Microcystis aeruginosa* (spring blooms), *Gomposphaeria lacustris,*

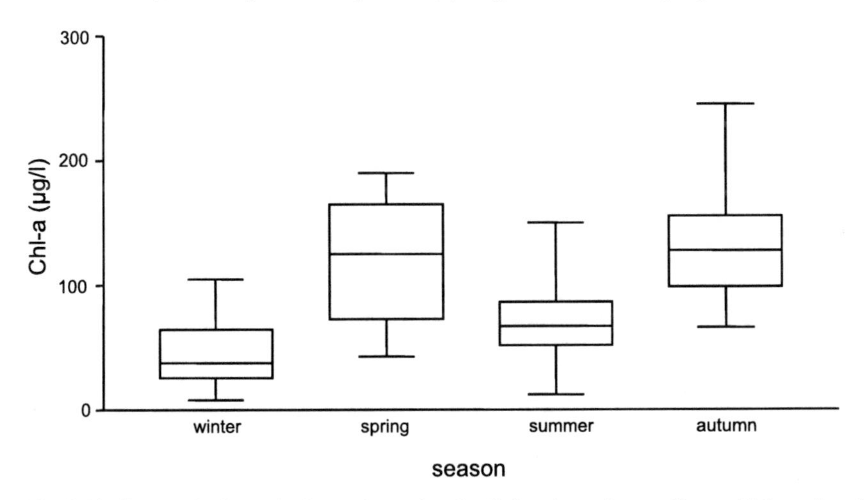

Fig.4.10. Seasonal phytoplankton dynamics for Lake Amstelmeer. Box-whisker plots for the period 1990–2000. Data from Hoogheemraadschap Hollands Noorderkwartier, unpublished. See Fig. 4.3 for comparison

Anabaena spec. (in July) and *Aphanizomenon floss-aqua* (August blooms). The algal densities are high. A clear water phase is not reached, not even in winter: On February 5th 1996, a concentration of 75 µg/l chlorophyll-a was measured under a covering of ice (*Oscillatoria spec.*) (Hogenbirk 1996).

Zooplankton

Only a limited amount of data is available pertaining to the zooplankton in the lake. In March–June 1996, the seasonal succession of the zooplankton was recorded (Hogenbirk 1996). The zooplankton was dominated by rotifers in March, while in April the dominance shifted to copepods. Cladocerans (viz. *Daphnia longispina, Bosmina spp.*) were present from the end of May onwards.

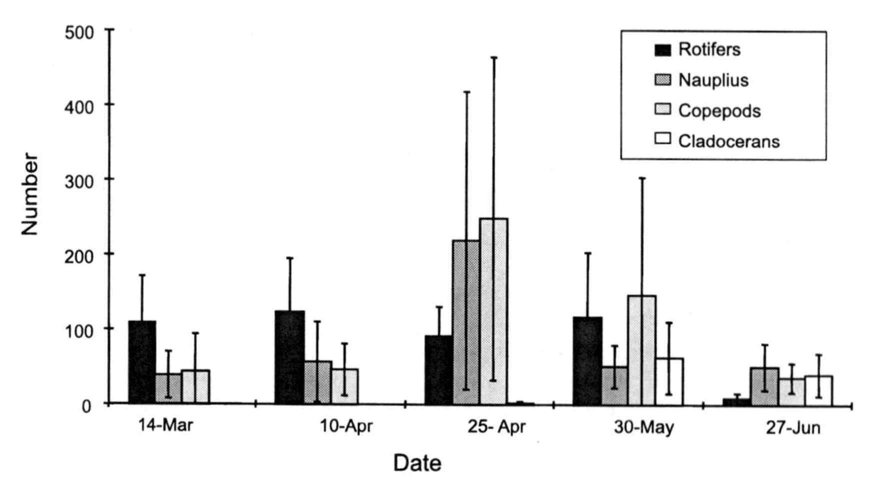

Fig. 4.11. Zooplankton community dynamics in the spring of 1996. Dates are average and standard deviations of 14 sampling sites in lake Amstelmeer (from: Hogenbirk 1996)

Resilience of the Plankton Community to Eutrophication

As for Lake Geestmerambacht, three experiments were carried out in order to test the effects of nutrient additions on the natural plankton community of the lake. The results are shown in Fig. 4.12. There was an intense response to the added nutrients by the phytoplankton.The addition of phosphorus resulted in very high chlorophyll-a concentrations (> 500 µg/l), especially in May. The zooplankton was clearly unable to control the development of the phytoplankton. This might, however, have been due to the high initial algal density.

Synopsis

Lake Amstelmeer has a continuous bloom of bluegreen and green algae throughout the year and no clear water phase. Algal densities remain high, even in winter. The reason for this is not the absence of daphnids, as daphnids are present in the

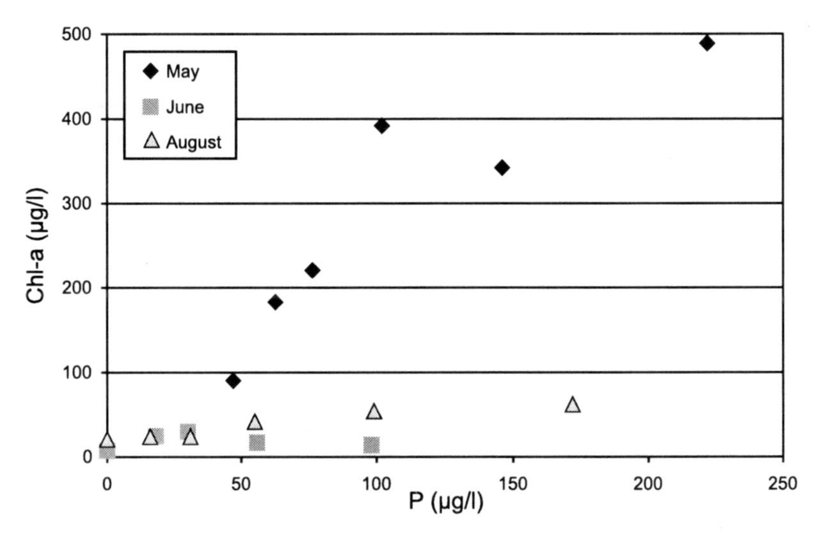

Fig. 4.12. Response of the phytoplankton (expressed as maximum chlorophyll-a concentration measured) after addition of nutrients to natural water from Lake Amstelmeer in indoor microcosms. The experiments were carried out in 1995. Initial zooplankton densities (cladocerans – copepods, in numbers per litre) were 2–9 in the May experiment; 1–5 in the June experiment, and 2–1 in the August experiment. The nutrient concentration on the x-axis is total phosphorus (background P + aditional P). See Fig. 4.8 for comparison

lake from May onwards. The robust response of algae to added nutrients in a spring microcosm experiment with natural water, suggested that the grazing by daphnids was suboptimal.

4.4 What Can Be Learned from These Lakes?

This chapter shows that two lakes with comparable nutrient levels can show very different phytoplankton dynamics. In Lake Geestmerambacht, the spring bloom was followed by a clear water phase during which the top-down control by zooplankton is very strong. Even a nutrient pulse will not lead to eutrophication phenomena during this period. On the other hand, Lake Amstelmeer shows a continuous algae bloom despite the presence of daphnids. The intense response to a nutrient pulse indicates that thegrazing capacity of the daphnid population is reduced.

In order to test this hypothesis, an experiment was executed with water from both lakes, at a point in time in which the phytoplankton communities resembled one another (Fig. 4.13). The natural zooplankton was removed and a standard community of cladocerans (*Symocephalus, Daphnia magna, D. longispina*) was added to the water. The phytoplankton community development was followed over time. At the same time, experimental systems without zooplankton were observed. The results are shown in Fig. 4.14.

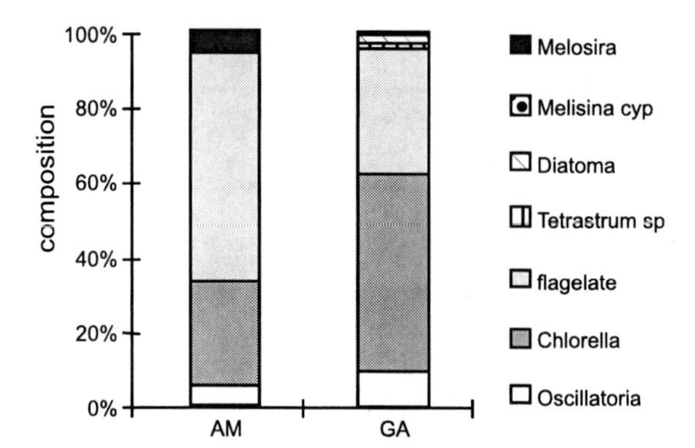

Fig. 4.13. Composition of the phytoplankton community at the start of the experiment. AM = Lake Amstelmeer; GA = Lake Geestmerambacht

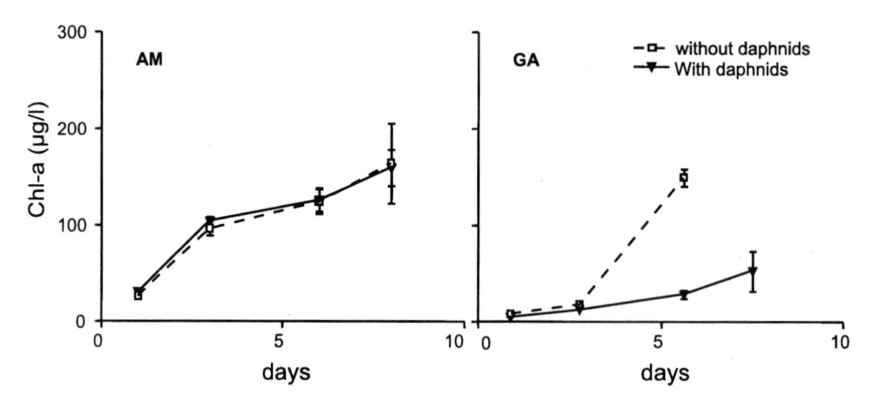

Fig. 4.14. Algal development in filtered Lake Amstelmeer water (left) and Lake Geestmerambacht water (right) with and without cladocerans

In the water from Lake Geestmerambacht, daphnid grazing resulted in a reduced chlorophyll-a density in comparison to the situation without daphnids. The grazing efficacy was 44%. In the Lake Amstelmeer water, however, a similar cladoceran community was completely unable to control the phytoplankton (grazing effectiveness of only 4%). The reproduction of daphnids (esp. *D. longispina*) resulted in an increase to the initial density of 9 individuals to a density of up to 60 per litre in the Geestmerambacht systems (see Fig. 4.15). In the Amstelmeer systems with daphnids, the final population density was 30 individuals per litre of water, which indicated some reproduction despite the extremely low grazing effectiveness.

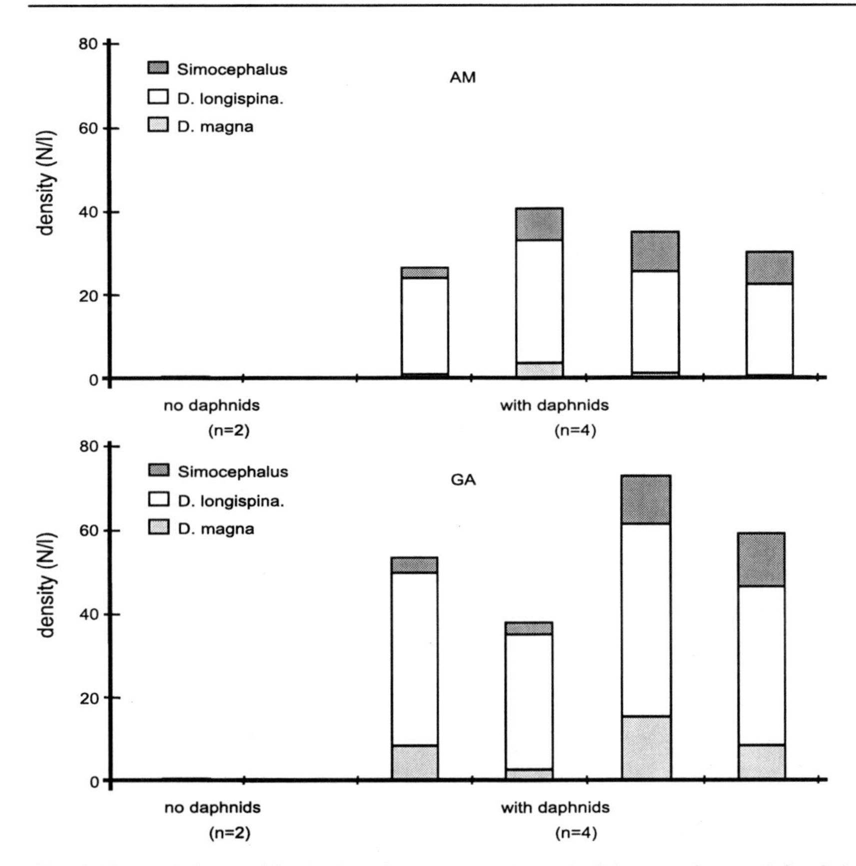

Fig. 4.15. Daphnid densities in the microcosms at the end of the experiments (after 8 days). Top: Lake Amstelmeer (AM); Bottom: Lake Geestmerambacht (GA). The initial density in the systems "with daphnids" was 9 per litre

Water Quality Is Critical

The conclusion from this experiment – where the zooplankton community is identical and the phytoplankton community almost identical – is that the reason for the significantly different grazing efficacies (44% and 4% respectively) must lie in the water quality. The water from Lake Amstelmeer was less suited to cladocerans than was the water from Lake Geestmerambacht. Reproduction could take place (which must be the case, as cladocerans are also found in the field situation) but the grazing effectiveness was reduced.

The factor responsible for this sub-optimal water quality is not easily identified (see Table 4.1). Lake Amstelmeer is influenced by many (potential) sources of pollution, such as drainage water from the surrounding agricultural areas, sewage treatment plant effluent and the dumping of dredged sediments. A recent inventory showed the presence of agricultural pesticedes in concentrations exceeding the prescribed quality standards in the canals transporting water to and from the lake (Van der Helm, 2000).

Table 4.1. Factors that may explain the difference in grazing effectiveness between the lakes

	Lake Amstelmeer	Lake Geestmerambacht
Water quality		
pH	8.5	8.6
Kjeldahl-N (mg N/l)	2.0	1.4
Total-P (mg P/l)	0.4	0.4
Salinity	0.5 – 1.5 ‰	0.25 ‰
Characteristics		
Isolation	part of a canal system; lake has a water transport and storage function	relatively isolated; used incidentally for water storage (flood control)
Land use	flower bulb culture, arable land	arable land, pastures (cattle, sheep)
Pollution sources	dumped polluted dredging materials (1982), lake received effluent of sewage treatment plant until 1996	some influence from recreation (swimming, surfing, diving, fishing)

Another factor lie in the fact that the lake is brackish, with a salinity of approx. 500–700 mg/l Cl (Hogenbirk 1996). Highest concentrations are approx. 900 mg/l in November; and concentrations are lowest in April (< 500 mg/l). In the deepest parts of the lake, the chlorine concentration can reach up to 1600 mg/l near the bottom (Anonymous 1994). Salinity is not very toxic to *Daphnia magna* (EC_{50} 48 h. for artificial sea salt 5600 mg/l; Grootelaar and Maas-Diepeveen 1988). In eco-assays, a LOEC for daphnid grazing was observed at 3000 mg/l (see Sect. 3.3), but some effects on grazing were observed at concentrations as low as 1600 mg/l (see Chap. 3). Other daphnids, such as *Ceriodaphnia dubia*, are more sensitive (EC_{50} 48h. 1189 mg/l Cl; Mount et al. 1997). This indicates that the highest concentrations in lake Amstelmeer could have affected the cladoceran community. As the maximum salinity is measured in November and it decreases during the spring, salinity could well explain the delayed cladoceran development.

An Incidental Case?

A similar difference in daphnid grazing effectiveness was observed by Madveev et al. (1994) in a comparison of two eutrophic Australian lakes: one in a dry forested area (Lake Dartmouth) and one in an agricultural area (Lake Hume).

The mean chlorophyll-a level in Lake Hume was high in comparison with Lake Dartmouth. In a bio-assay, added zooplankton had no clear effect on the algal density in water from Lake Hume. The algal concentration decreased with increasing daphnid density in water taken from Lake Darthmouth, indicating effective grazing. Nutrient enrichment resulted in enhanced algal density in Lake Hume, while in Lake Dartmouth this did not result in enhanced algal density due to daphnid grazing control (Fig. 4.16).

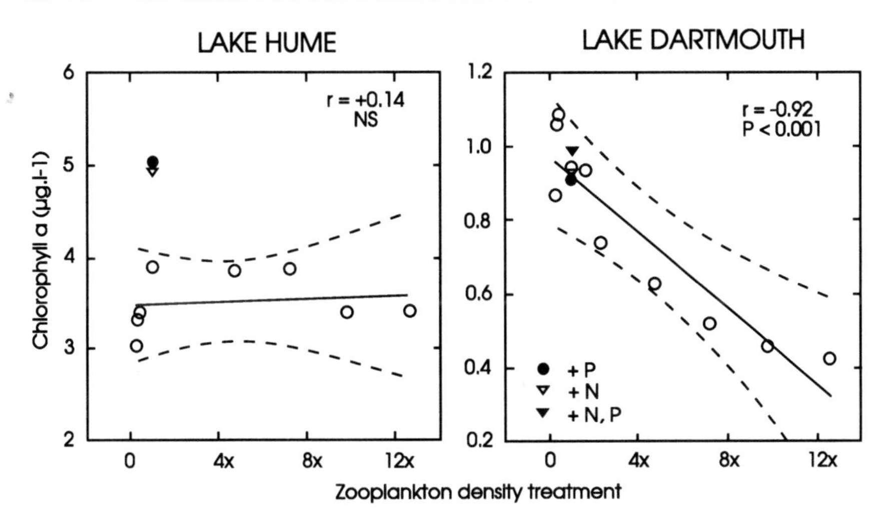

Fig. 4.16. Phytoplankton concentration (chlorophyll-a) as a function of zooplankton density in two Australian lakes. In lake Dartmouth (right panel), increasing zooplankton density reduced the algae density (grazing). In lake Hume, no grazing seemed to take place (left panel). From Matveev et al. 1994

The main zooplankton species in Lake Dartmouth was *Daphnia carinata*, which was almost completely absent in Lake Hume, where other cladoceran species and copepods dominated. Although the authors do not link these observations to differences in toxicant stress, the land use suggests that the pesticide concentration in Lake Hume may well have been higher than in Lake Dartmouth. Matveev and Madveeva (1997) estimated a grazing effectiveness for a cladoceran community dominated by *D. carinata* at up to 0.80 per day^{-1}, which is important for the development of a clear water phase. Significant grazing is predicted when the cladoceran / phytoplankton biomass ratio is greater than > 0.1.

5 New Perspectives for Eutrophication Management

5.1 A New Dimension in Lake Eutrophication Management

Abiotic Stressors and Toxic Chemicals: Factors to Be Considered for Lake Restoration

Restoration management aims at reducing nutrient inputs and favouring daphnids in order to control phytoplankton, thereby reducing turbidity by turning the system from a turbid into a clear state (Carpenter et al. 1995a). This is especially important in spring, when environmental conditions are optimal for a stimulation of the phytoplankton biomass. However, restoration measures have proven to be ineffective over the long term in many cases, possibly because of reduced grazing effectiveness by daphnids on algae. Ineffective daphnid grazing, due to toxic substances and other abiotic stressors, has been proven in many laboratory and mesocosm studies to occur. Therefore, it may be worthwhile considering the presence of toxic or other stress factors in the waters where biomanipulation has failed in comparison to the waters in which it was successful. Toxicants can act as one of the forward switches responsible for a shift from clear to eutrophied water (see Fig. 5.1). The crucial question however is: does this occur in the field situation?

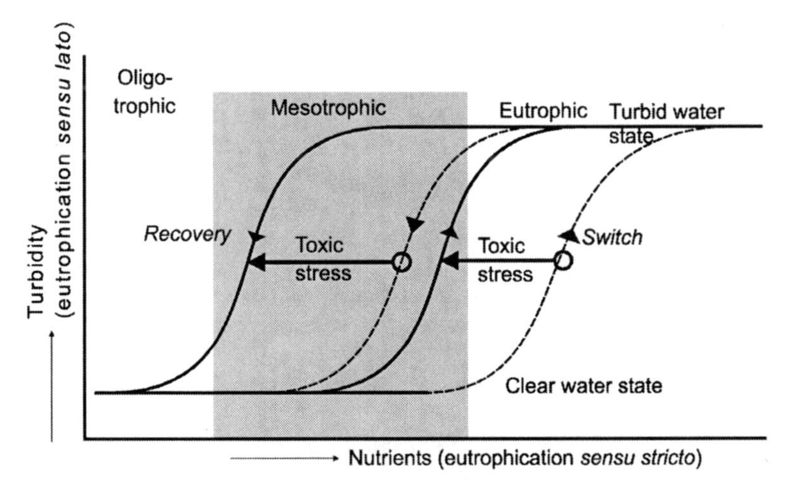

Fig. 5.1. Hysteresis effect in the response of lake turbidity to changes in nutrient status (adapted from Scheffer et al. 1993), extended for the impact of toxic stress

Table 5.1. LOEC data for toxic anorexia in daphnids from the toxicants tested in Chap. 3, in comparison to environmental concentrations and Dutch water quality standards for surface water

Substance	Source	LOEC daphnid grazing[1] (µg/l)	Concentration range in sewage water (µg/l)[6]	Concentration range in WWTP effluent (µg/l)[6]	Background concentration (µg/l)[2]	MPC (µg/l)
Cadmium	Mainly metal coatings	100	≤0.52	≤0.08	≈0.08[3]	0.4
Copper	Drinking water pipes, algicides, herbicides	30	76	≤10	≈0.44[3]	1.5
Dimethoate	Pesticides	18–320	≤0.27	≤0.01	≈0.05[5]	23
Fluoranthene	Treated wood, antifouling coatings, pesticides	5.6–10	≤0.27	≤0.01	≈0.009[4]	0.5
NaCl	Marine influence	3200000	–	–	–	–

MPC: Maximum Permissible Concentration (MTR in Dutch) n.s: not significant, n.a: not applicable
[1] see Table 3.3
[2] 90 percentile for fresh water systems; 10% of the measured concentrations were higher in 1998
[3] De Bruijn et al. 1999
[4] Van Steenwijk and Mol 1996
[5] Teunissen-Ordelman and Schrap 1996
[6] Gommers et al. 1999

Is Toxic Anorexia Likely to Occur in Practise?

LOEC data for toxic anorexia from some toxicants derived from the experiments in Chap. 3 are reviewed in Table 5.1. The LOEC is compared to concentrations of the selected substances in sewage water and effluent from a waste water treatment plant. Discharge of untreated sewage water may yield concentrations sufficiently high in order to stimulate occurrences of toxic anorexia, due to e.g. copper. The pesticide concentration in agricultural polder ditches can incidentally exceed the LOECs for toxic anorexia.

A further comparison is made with water quality standards. From the table, it can be concluded that, for most of the substances, the LOEC for toxic anorexia is well below the quality standard used for surface waters in the Netherlands. These MPC's (Maximum Permissible Concentrations) are calculated from the principle

Table 5.2. Highest concentration of pesticides in regional waters in the Netherlands, in comparison to the quality standard (MPC). Only substances that exceed the standard with a factor 10 or higher have been selected. (i) insecticide; (f) fungicide; (h) herbicide; (n) nematicide. From Teunissen-Ordelman and Schrap (1996)

More than 10000 * MPC	**1000 – 10000 * MPC**	**100 – 1000 * MPC**	**10 – 100 * MPC**
Dichloorvos (i)	Parathion (i)	Carbendazim (f)	Benomyl (f)
		Pirimicarb (i)	Diuron (h)
		Lindane (i)	Oxamyl (n/i)
		Mevinfos (i)	Mecoprop (h)
			Atrazin (h)
			MCPA (h)
			Simazin (h)
			Captan (f)
			Terbutylazin (h)
			Diazinon (i)
			Dinoseb (h)

that 95% of the species in the ecosystem must be protected. One exception, however, is dimethoate, where the LOEC for toxic anorexia (18 µg/l) is comparable with the quality standard (23 µg/l).

The conclusion can be drawn that – with the exception of dimethoate – the Dutch quality standards will prevent toxic anorexia from occurring in practise. However, it is important to realize that quality standards are not always met in practice, and that field concentrations of substances such as metals and agricultural pesticides may exceed the quality standards (see Table 5.2). Therefore, toxic anorexia is a factor worthy of consideration in local situations where concentrations exceeding quality standards are likely to occur.

In extreme cases, the MPC can be exceeded to a factor of more than 10,000. However, concentrations of more than 10 or 100 times the quality standard are measured for many agricultural pesticides (see Table 5.2).

An Example: Toxic Anorexia in Ditches in a Flower Bulb Cultivation Area

High pesticide concentrations may occur in regional waters. An example of such a regional situation occurs in the northwestern part of the Netherlands, in the Province of Noord-Holland. In this polder area, cultivation of flower bulbs (e.g. tulips, sword lilies) takes place on a large scale. Figure 5.2 mpression of the land usage in this area. In 1996, an eco-essay was performed with the polder water and cultivated daphnids in order to test the daphnid grazing effectiveness in several locations in the flower bulb cultivation area (see Fig. 5.3).

Significant differences in daphnid grazing effectiveness can be found between different locations in this area. The grazing effectiveness is expressed as CR*Cf (see Chap. 2), which means that high values (b, c) indicate efficient grazing of algae by daphnids, and low values (a, d, e, f) inefficient grazing. Locations a, d and e are situated in the middle of an extensive bulb cultivation area. Location f is located in the middle of a horticultural area with greenhouses and some flower bulb

Fig. 5.2. Flower bulb cultivation (tulips) in the Netherlands

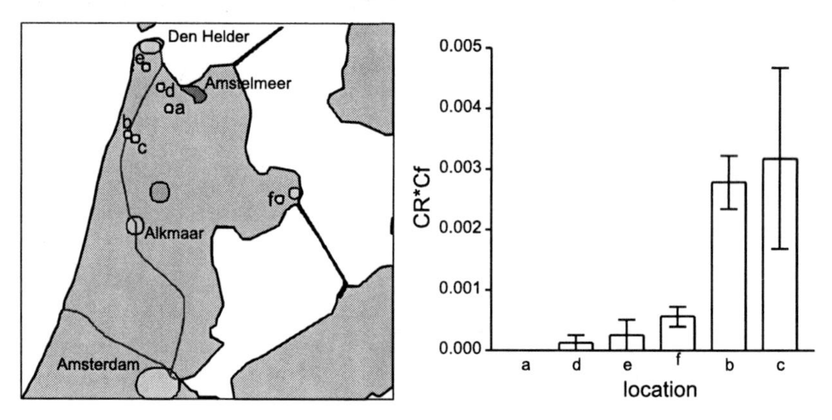

Fig. 5.3. Daphnid grazing effectiveness in a daphnid community exposed to polder water from 6 locations (a–f) in a flower bulb cultivation area in the Netherlands (initial density per litre: *Ceriodaphnia* 3, *Daphnia magna* 3 and *Simocephalus vetulus* 3)

cultivation. Locations b and c are at the southern edge of the bulb cultivation area with large areas of pasture with cattle and sheep. This suggests a correlation of daphnid grazing reduction with pesticide loadings.

The main water flow in the northwestern region (all locations, except f) is from south to north.

In 1996, pesticides were measured at these six locations by the regional water board (Hoogheemraadschap Uitwaterende Sluizen). Pesticide concentrations (esp. carbendazim) were relatively high at locations a, d and e; and low at locations b and c; thus confirming the suggestion that daphnid grazing reduction may be related to the pesticide load.

A study in this area in 1998 (Van der Helm, 2000), in which pesticide concentration were measured from March to November at 12 different locations, showed a decrease in water quality in the flow direction from south to north, i.e. a decrease in water quality from the area where b and c are located towards the area where a, d and e are located. The main substances responsible for this decreasing water quality were carbendazim and metalaxyl.

Conclusion: Toxic Anorexia, a Factor to Be Considerd?

In Chaps. 2 and 3, toxic anorexia has been demonstrated as a mechanism that can act as a forward switch, inducing a shift from a clear water state towards a eutrophied state in small, artificial water bodies (80 laboratory cultures, and 5 m^3 outdoor mesocosms). The concentrations of the toxic substances that induce this shift (such as metals and pesticides) are generally lower than the value at which toxic effects such as mortality or growth reduction occur. Despite this, the LOECs for toxic anorexia are sufficiently high for environmental quality standards to be safe. Thus, toxic anorexia is not expected to occur on a large scale in natural waters. In specific regional waters, however, environmental quality standards may be exceeded. An example is areas with intensive agriculture, where pesticide concentrations as high as 100 times the prescribed quality standard or more are observed. Toxic anorexia may well occur at these concentrations. It was demonstrated for an agricultural area intensively used for flower bulb cultivation that the grazing effectiveness was very low at several sites where a high level of pesticides was likely to have been present. Therefore, incidental pollution, resulting in high concentrations of substances such as pesticides, must be considered as a risk factor for toxicant-induced eutrophication phenomena. The fact that pesticide (e.g. insecticide) use coincides with the seasonal algal bloom makes this period extra critical (Fig. 5.4).

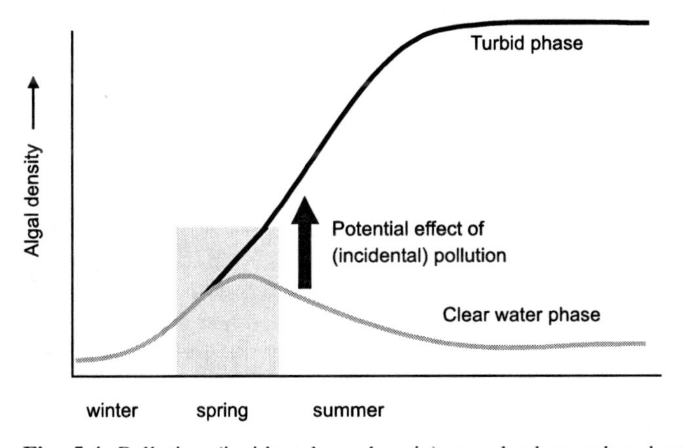

Fig. 5.4. Pollution (incidental or chronic) may lead to reduced grazing effectiveness by daphnids. When this happens in the critical season for plankton development (spring), the pollution event may act as a forward switch, forcing the system into a turbid state. Thus, the effect of the pollution may last for much longer than the pollution itself

When considering the causes of eutrophication for a specific water body, or the strategy for the restoration of this water body, toxicants (incidental pollution in the present or the past) must be considered as a potential forward switch.

5.2 How to Address Ecotoxicology in Eutrophication Management

Present Lake Restoration Strategies

Lake restoration strategies are primarily aimed at reducing the external phosphorus loading. However, once the external loading has been reduced, recovery of the lake may be inhibited by an internal phosphorus load in the sediments, in which most of the phosphorus has accumulated (Hosper 1997; Scheffer; 1998). In the past decennia, lake restoration has developed from processes of trial-and-error towards an independent scientific discipline. The restoration strategy for eutrophicated waters generally consists of seven steps (Moss et al. 1996a; Moss 1998):

1. Problem diagnosis and target setting
2. Removal of existing or potential forward switches
3. Reduction of nutrient load
4. Biomanipulation
5. Re-establishment of plants
6. Re-establishment of appropriate fish community
7. Monitoring of the results

1. Problem Diagnosis and Target Setting
The first step in eutrophication management is to make a clear analysis of the problems to be resolved with restoration measures. The target can be the re-establishment of the former biodiversity as deduced from historical evidence, to create an attractive, though not a necessarily diverse community for general amenity, to establish clear water for abstraction and less expensive subsequent treatment, or a robust general recreational fishery.

It is important to gather background information on any past changes in the conditions of the detoriated water body; which would indicate that any forward switches need to be identified. Forward switches may include mechanical damage, boat damage, exotic grazers, herbicides used to remove weeds, water level increase, salinity increase, pesticides and selective fish kills. The effects that pesticides may have had on *Daphnia* and other crustacean grazers are perhaps the most difficult to discover. Some decades ago, which was a time when the use of pesticides was much more liberal, , these effects were not fully realised. The evidence for their importance is based on the effects of recorded spills and on the detection of residues in sediment cores, coincident with changes in the composition of the zooplankton community.

2. Removing Forward Switches
A restoration programme may be blocked if forward switches that cannot be removed, or seriously reduced in influence, are present. The effectiveness of a switch may be reduced if the nutrient loads can be lowered. Some forward switches are difficult to remove at all, but nevertheless, some obvious switches can be removed and this must be combined with nutrient control.

3. Reduction of Nutrient Load
The development of increasing crops of algae requires phosphorus and nitrogen to be supplied in a ratio of about 1 to 10. Growth can be greatly reduced by severely restricting the supply of only one nutrient. It is far easier to control phosphorus rather than nitrogen in external sources. Phosphates can be easily precipitated and can usually be traced to a few concentrated point sources. Phosphorus can accumulate in lake sediment, in contrast with nitrogen. Nitrates and ammonium are very soluble, cannot easily be precipitated and are emitted from a number of diffusive sources. Reduction programmes are most successful when internal loads are first removed before dealing with external loads.

4. Biomanipulation
The most effective biomanipulations consisted of the removal of the entire planktivorous fish community, combined with the introduction of piscivores to deal with any residual fish which escaped netting or electrofishing. If carried out efficiently, this process almost always results in an increase in daphnia populations and clearing of the water. Failures are usually due to insufficient fish removal or because invertebrate daphnia predators have increased in number.

5. Re-establishment of Plants
Re-establishment of plants does not automatically occur after biomanipulation. Seeds or vegetative propagules of former overwintering plants may not be sufficiently viable for the stimulation of the growth of a new plant community. The sediments laid down under phytoplankton communities are often soft and amorphous and may not provide adequate rooting conditions. Therefore, plants may need to be introduced and provided with protection against physical disturbance and birds, which can proove costly.

6. Re-establishment of Appropriate Fish Community
Once the plants are well established, the fish community must be reconstructed. The ultimate aim of restoration is the re-establishment of a self-sustaining ecosystem. If a suitable fish community is not artificially established, one may develop spontaneously from invasion and from residual fish, which may disturb the plant community and lead to a return to the turbid state.

7. Monitoring of the Results
In the long term, few restoration attempts are entirely successful. Some of the reasons for this are understood, such as failure to remove or sufficiently reduce forward switches, nutrient reduction or fish removal. The new situation will always be potentially unstable if the restoration of the lake is not accompanied by sustain-

able management of the catchment area, with special attention paid to the use and emission of fertilizers and pesticides.

Inclusion of Ecotoxicological Aspects in These Strategies

The inhibition of daphid grazers by toxic substances or other abiotic stressors is not addressed explicitly in current restoration practise. However, as we have seen in Chap. 3, toxic anorexia may well be an important factor in specific cases. By explicitly addressing the possibility of toxic anorexia occurring during lake restoration procedures, the risk of projects being unsuccessful due to the influence of abiotic stressors being overlooked is reduced. An approach is proposed in Fig. 5.5.

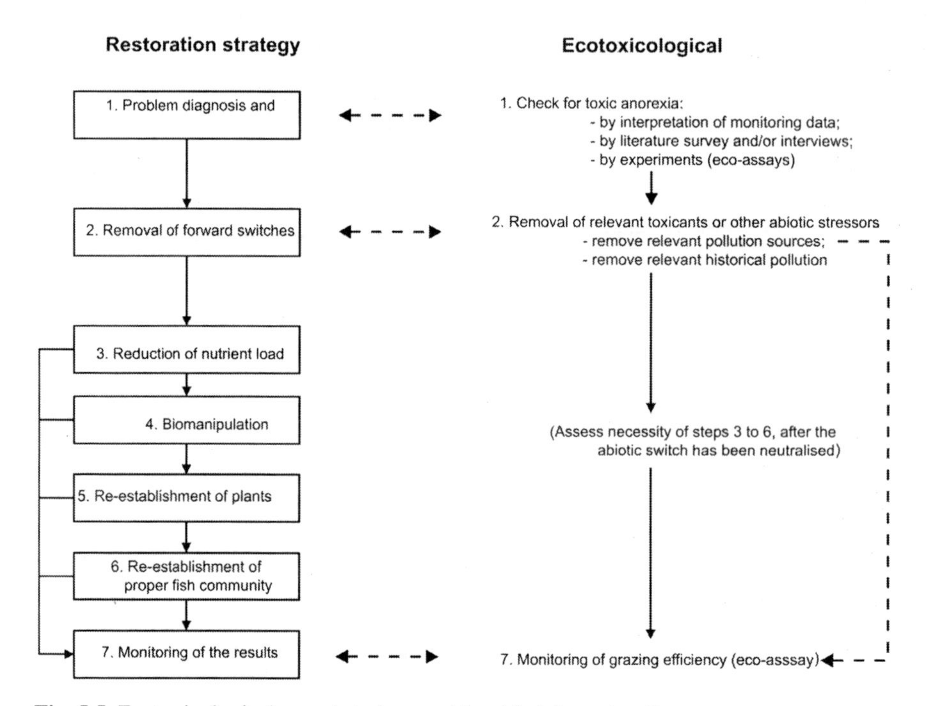

Fig. 5.5. Ecotoxicological aspects to be considered in lake restoration

Step 1 – Detection of Toxic Anorexia

It is of the utmost importance to address the possibility of toxic anorexia from the start of the restoration project. During diagnosis of the problem and the target setting phase, the possibility that zooplankton grazing may be inhibited by abiotic stressors must be assessed. Basically, there are three possible approaches (see Table 5.3). The first is to deduce, from data on zooplankton and phytoplankton in the lake, the possibility of reduced top-down control. The second is to assess the possibility of the exposure of the lake to toxic substances (or other abnormal abiotic conditions), either at present time or in the past. The third and most direct approach is to measure the grazing efficiency of the zooplankton community by executing a plankton eco-assay experiment. This may also be a wise approach for verifying the findings from the other two approaches.

Step 2 – Removal of Relevant Toxicants

When there is serious reason to assume that toxic inhibition of algal grazing plays a role in the eutrophication of a lake, methods must be considered for the removal or reduction of the source of the pollution. This may either be pollution from the past, which is still present as sediment contamination (such as the use of DDT as an agricultural pesticide, Stansfield et al. 1989), or still on-going pollution (e.g., discharge of STP effluent). In the case of historical pollution, several (cost-effective) options may be considered as alternatives to the removal of the contaminated sediment (see Table 5.4). It is important to develop a strategy for the removal of the toxicants (i.e. which substances must be removed? Which residual level is acceptable?). Toxicity Identification and Evaluation (TIE) techniques are being developed in order to identify the substances that cause toxic effects in environmental samples (Huwer and Brils 1999).

Step 7 – Monitoring

After appropriate measures have been taken, the success of the restoration project must be monitored. As described in step 1, the most direct method for measuring the grazing efficiency (top-down control) is to execute a plankton eco-assay.

Table 5.3. Methods to detect toxic anorexia (present and past) in lakes

How to detect toxic anorexia?		
1. Interpretation of (existing) field monitoring data	–	Daphnids grazing factor (ratio Chlorophyl a : daphnids). A low value is an indication of a reduced top-down control of the algal density
	–	Daphnid grazing effectiveness. A low value indicates ineffective grazing by Daphnia.
	–	Daphnid dominance: Daphnid: Copepod ratio. A ratio < 1 indicates unfavorable conditions for Daphnids competing with Copepods, resulting in an ineffective top-down control of phytoplankton.
2. (Historical) survey	–	The presence of toxicants (e.g., monitoring data of the local water management board) may indicate possibility of toxic anorexia
	–	Other conditions that may indicate toxic anorexia are high salinity, high or low pH, or other abnormal abiotic conditions
	–	Historical pollution may indicate a toxicant-induced switch in the past. Literature or interviews may yield information on historical discharges, land use, industry, etc.
3. Experiments	–	The most direct approach for detecting a reduced top-down control due to toxic substances or abiotic stressors is to execute a plankton eco-assay, as described in Chap. 2. This is a direct measurement of the zooplankton grazing efficiency.

Table 5.4. *In-situ* methods for reducing the effects of (historical) sediment pollution (from Van der Gun and Joziasse 1999). Potentially applicable methods: not all measures have been proven in practice

Approach	Contaminants	Technological concept
Stimulate microbiological degradation	Organic (PAH's, oil)	– Addition of electron-acceptors; – Addition of suitable microorganisms; – Addition of nutrients; – Change local environmental conditions
Stimulate microbiological reductive dechlorination	Chlorinated organic compounds	– Change local environmental conditions
Stimulate biological degradation	Organic	– Degradation of substances by plants
Stimulate biological concentration and removal	Metals (esp. Ni, Zn), phosphates	– Phytoextraction: concentration in plants
Mobilisation	All	– Mobilise contaminants and treat water phase
Chemical reactions	All	– Chemical oxidation or reduction of chemicals
Immobilisation of contaminants	Metals	– Precipitation as hydroxide/ complex – Phytostabilisation – Precipitation as sulfide – Bind to inorganic matrix – Adsorption to clay/ organic matter
	Organic	– Adsorption to clay/ organic matter
	Phosphorus	– Precipitation of Fe-phosphorus
Dredging and disposal of contaminated sediments	All	– Removal of source
Reduce advective transport	All	– Increase hydraulic resistance – Hydrological isolation

Using Plankton Eco-assays

A good and cost-effective way to check for toxic anorexia is to execute a plankton eco-assay. This method is described in more detail in this section.

The concept of the plankton eco-assay facilitates the assessment of the grazing effectiveness of a daphnid population under various conditions and has wide applicability. The actual test design can be adapted to meet the specific goals of the test. Thus far, the plankton eco-assay has been used successfully for the determination of the effect concentration of various toxicants (see Chap. 3). It has also served as a range finding test for mesocosm (pond) studies with pesticides (Foekema et al. unpublished data) and as a test for determining the impact of surface water quality on the daphnid community (see Chap. 4).

Despite the wide applicability of the test, the basis is always similar. A plankton eco-assay is performed in 50–100 litre containers, with a height: width ratio of 2–3. The test containers are filled with the water to be tested and provided with continuous gentle aeration. For ca. 16 hours per day, the systems are illuminated by a light source suitable for algal development. When the light source is very close to the water surface, 'overheating' of the water must be prevented by, for instance, working in a temperature controlled, cool environment. The measurement of some parameters that can be critical to daphnids (e.g. dissolved oxygen, pH, NH_3, NO_2) increases the reliability of the test results.

During the test, the development of the algal community in the test containers is followed by measuring the chlorophyll-a concentration. In order to determine the grazing effectiveness of the daphnid population, a reference container with the same water type and algal community, but without daphnids, is tested in parallel.

When the algal development in the containers without daphnids is no longer exponential (usually within 10 days), the test is terminated and the zooplankton density is determined.

More information on how to interpret the results of an eco-assay (i.e. calculation of grazing effectiveness) can be found in Sect. 2.2 of this book.

When the eco-assay is applied for the evaluation of surface water quality, two experimental approaches can be followed: either to determine the grazing effectiveness of the resident plankton community or to test the suitability of the water for the development of an effective daphnid community.

For the first approach, no cultivated daphnids are added to the test containers and the development of the resident zooplankton community is followed over a series of nutrient loads. The collection of the water sample is critical and must be done in such a way that the natural zooplankton community is undamaged. The use of buckets for collection of the water is recommended. If for some reason this handcraft method is not suitable for the local situation, but a non-centrifugal pump can be used. It should be realised that part of the daphnid community might avoid the water currents produced by the pump, resulting in daphnid densities that are too low in the collected water, and therefore would be an underestimation of the potential grazing effectiveness.

The second approach is more suitable for the determination of the potential impact of water quality on daphnid grazing and subsequently on the algal development in surface waters. The zooplankton is removed by sieving the water sampled in the field. Depending on the nutrition status of the water, nutrients can be added in order to ensure hypertrophic conditions and, therefore, the potential for exponential growth in the resident algal community. The N : P ratio should be 1:16 on molar bases (c.f. the Redfield ratio) with a P-content between 0.5 and 1 mg per litre at the start of the test.

The critical initial daphnid density for the water type can be determined by supplying test containers with various initial daphnid densities.

The methods described in Chap. 2 enable a comparison of the daphnid grazing effectiveness in various water types/algal assemblages to be made. Determination of the dominant algal species at the beginning and the end of the test is useful for the interpretation of the results, as in some cases the low performance of the daphnid population is the result of an inedible (e.g. filamentous) algal community. This type of testing is preferably performed with a well-defined single daphnid species of a certain age, or a well-defined mixture of daphnid species, in order to allow for bench-marking.

Eco-assays with the aim of determining the effect concentration of a substance (e.g. to identify which toxicant might be responsible for observed toxicity in field samples) can best be performed with an artificial medium and cultivated algae in order to allow standardisation. The test medium should contain high nutrient concentrations and be suitable for daphnid development. Satisfactory results have been achieved with the OECD algal growth medium (OECD 1993). Algae from an exponentially growing culture are added until the chlorophyll-a concentration is 5–10 μg/l.

Test containers with and without daphnids are tested in parallel over the same concentration series of the test substance. On account of the use of a concentration series, the results can be expressed on a relative basis as GR%, which can be calculated according to the model presented in Chap. 2.

A well-designed test over a concentration series will give information about the effect of the test substance on 3 levels: algal development (algal growth rate), daphnid development (daphnid reproduction) and zooplankton–phytoplankton interactions (grazing effectiveness). The plankton eco-assay test can therefore be regarded as an efficient test for the determination of the potential impact of a particular substance on a plankton community.

5.3 Epilogue

This book provides a reflection of experiments carried out at the TNO laboratories in Den Helder, The Netherlands, during the 1994–1999 period. During experiments on the interacting effects of nutrients and toxicants on plankton communities, the principle of toxic anorexia emerged and was proven to occur in practice many times. With this book, the resultant principles of the response of plankton communities to toxicants in eutrophied (*sensu stricto*) waters are made available to limnologists, ecotoxicologists and, especially, water managers working in the area of eutrophication management or lake restoration.

The first main conclusion that can be drawn is that the paradigm that the algal density is a linear function of the limiting nutrient (phosphorus in most cases) is an oversimplification. This view, which is reflected in the Vollenweider models (OECD 1982), certainly has its use in predicting the potential maximum algal density at a given nutrient loading, but is not accurate in predicting the actual algal concentration in a specific lake. The actual algal density is a function of two equally important factors: the nutrient level (bottom-up control) and the grazing capacity of the zooplankton (esp. Daphnid) community (top-down control). This new paradigm is illustrated by more recent studies, e.g. Portielje and Van der Molen (1997b).

The second main conclusion is that the top-down control can be reduced by toxic substances, at concentrations that are too low to have adverse effects on phytoplankton development and that are well below lethal concentrations for zooplankton. Toxicant exposure may lead to reduced grazing effectiveness by daphnids, the zooplankton group most capable of controlling algal growth. This "toxic anorexia" in daphnids becomes apparent as eutrophication sensu lato: high algal densities and turbid water. The principle has been demonstrated in 80 litre indoor systems (plankton eco-assay), outdoor mesocosms and outdoor plankton enclosures. There are several examples of lakes where toxic anorexia may have played a role. This has been demonstrated for toxic substances such as insecticides and metals and also for specific abiotic environmental conditions, such as salinity. The occurrence of toxic anorexia in field samples can be demonstrated with the standardised plankton eco-assay.

Finally, the third main conclusion is that toxic anorexia is a factor for consideration in lake management and lake restoration. Lake ecosystem response to nutrient loading is not a gradual, continuous process, but is characterised by two alternative stable equilibria: a clear water state and a turbid state, both with a high resistance to changes (Moss et al. 1996a; Scheffer et al. 1993). A shift from one stable state to the other is called a switch, and can be induced by an external factor. The conclusion drawn by the authors of this book is that toxic anorexia in daphnids is a factor to be considered as a possible cause for the occurrence of a switch from clear to turbid lakes. Although LOECs for toxic anorexia for most toxic substances are well above average concentrations found in pristine waters, switches may well be caused by pollution events which occurred in recent history. Contamination accumulated in sediments, and the loading of surface waters with pesticides, industrial effluents, sewage waters and effluents from waste water treatment plants can cause local incidental conditions which may result in toxic anorexia occurring amongst daphnids. Therefore, toxic anorexia should be taken into consideration when a strategy for eutrophication management or lake restoration is developed. The plankton eco-assay is a valuable analysis tool in this context. When toxic anorexia cannot be excluded, it is recommended that the toxicant levels are first reduced before nutrient reduction or biomanipulation measures are implemented. This will increase the chances of the restoration measures succeeding.

It must be stressed that although only the consequences of toxic anorexia amongst daphnids for phytoplankton dynamics and bloom are described, the disturbance of phytoplankton-zooplankton interactions may have consequences for the entire aquatic ecosystem, due to disturbance of the trophic cascade. This is an important topic for future research.

This book addresses toxic anorexia in temperate shallow lakes. However, similar principles may be found in other habitats, such as esturine or coastal waters. The principle of toxic anorexia may also be valid for other algae-grazer interactions. For example, Ostroumov (2002) hypothises toxic anorexia from top-down control of phytoplankton by filter-feeding bivalves (e.g. *Dreissena*). The authors hope that this book may stimulate other applied scientists to adopt this line of thought, thus combining limnology and ecotoxicology in order to acquire more understanding of the ecological effects of pollution. The authors also express the hope that it may be the basis for more effective results in lake restoration for those engaged in water management.

Acknowledgements

Many people have contributed to this book. Most of the experiments that form the foundation for this book were carried out at the TNO Laboratories in Den Helder, The Netherlands, under the supervision of Edwin Foekema and Robbert Jak. The authors would like to thank everyone who has contributed to these experiments: Gerrit Hoornsman, Liesbeth van der Vlies, Dennis van der Veen, Sherri Huwer, Wendy Koegelberg, Marijke van der Meer and Henk van het Groenewoud. This book would not have been possible without their continuous effort and enthusiasm.

The experimental work was carried out in close co-operation with the University of Alicante, Spain, and the Université de Savoie (ESIGEC). The authors extend their gratitude to Dr. Daniel Prats, Dr. Pilar Hernandez, Juan Carlos Asensi, Mari José Navarro, Laura Rull, Dr. Gérard Blake and Dr. Bernard Clement for their contributions and for the stimulating discussions.

The authors are much indepted to a group of experts who were willing to invest their time and knowledge in providing commentary on the first draft of this manuscript in a 3-day peer-review workshop (December 1999) in Callantsoog (the Netherlands): Donald Baird (University of Stirling, UK), Gérard Blake (ESIGEC, France), Bernard Clement (ENTPE, France), John Driver (Albright and Wilson, UK), Pilar Hernandez (University of Alicante, Spain), Brian Moss (University of Liverpool, UK), Sergei Oestromov (State University of Moscow, Russia), Hans Toni Ratte (Aachen University of Technology, Germany), Loreto Rossi (University of Rome “La Sapienza”, Italy), Nico van Straalen (Free University of Amsterdam, the Netherlands), Chris Thornton (CEFIC-CEEP, Belgium) and Frans Visser. Their contributions were invaluable to the focussing of this book.

Last but not least, the authors thank CEFIC – CEEP and the Dutch Ministry of Transport and Water Management (RWS-RIZA) for their financial support and Springer Verlag (Wim Salomons) for their continuous support and enthusiasm.

Martin Scholten, Research Coordinator

References

Allan JD (1976) Life history patterns in zooplankton. Am Nat 110:165–180

Andersen T (1997) Pelagic nutrient cycles. Herbivores as sources and sinks. Ecological Studies 129, Springer-Verlag, Berlin.

Andersson G, Granéli W, Stenson J (1988) The influence of animals on phosphorus cycling in lake ecosystems. Hydrobiologia 170:267–284

Anonymous (1994) Hydrobiologisch onderzoek Amstelmeerboezem. Hoogheemraadschap van de Uitwaterende Sluizen in Kennemerland en Westfriesland

AquaSense (1996) Bestrijding van eutrofiering in de Geestmerambachtplas. Deelonderzoek: inventarisatie fysische en biologische parameters tbv eutrofieringsonderzoek. Commissioned by: Waterloopkundig Laboratorium. Report number 96.0448

Baird DJ, Barber I, Bradley M, Soares AMVM, Calow P (1991) A comparative study of genotype sensitivity to acute toxic stress using clones of Daphnia magna Straus. Ecotoxicol Environ Saf 21:257–265

Bales M, Moss B, Phillips G, Irvine K, Stansfield J (1993) The changing ecosystem of a shallow, brackish lake, Hickling Broad, Norfolk, UK. II. Long-term trends in water chemistry and ecology and their implications for restoration of the lake. Freshwater Biol 29:141–165

Barnes RSK, Mann KH (1993) Fundamentals of aquatic ecology. Blackwell Scientific Publications. Oxford. 2nd ed

Baudo R (1987) Ecotoxicological testing with Daphnia. In: Peters RH, De Bernardi R (eds) Daphnia. Mem Ist Ital Idrobiol 45:461–482

Beklioglu M, Moss B (1995) The impact of pH on interactions among phytoplankton algae, zooplankton and perch (Perca fluviatilis) in a shallow, fertile lake. Freshwater Biol 33:497–509

Benndorf J (1990) Conditions for effective biomanipulation; conclusions derived from whole-lake experiments in Europe. Hydrobiologia 200/201:187–203

Bleiker W, Schanz F (1997) Light climate as the key factor controlling the spring dynamics of phytoplankton in Lake Zürich. Aquat Sci 59:135–157

Blockwell SJ, Taylor EJ, Jones I, Pascoe D (1998) The influence off fresh water pollutants and interaction with Asellus aquaticus (L) on the feeding activity of Gammarus pulex (L). Arch Environ Contam Toxicol 34:41–47

Burns CW, Schallenberg M (1998) Impacts of nutrients and zooplankton on the microbial food web of an ultra-oligotrophic lake. J Plankton Res 20:1501–1525

Burns CW (1968) The relationship between body size of filter-feeding cladocera and the maximum size of particle ingested. Limnol Oceanogr 13:675–678

Butler NM, Suttle CA, Neill WE (1989) Discrimination by freshwater zooplankton between single algal cells differing in nutritional status. Oecologia 78:368–372

Cairns J, Cherry DS (1993) Freshwater multi-species test systems. In: Calow P (ed) Handbook of ecotoxicology. Vol 1. Blackwell Scientific Publications, Oxford. pp. 101–116

Carpenter SR, Kitchell JF (1993) Experimental lakes, manipulations and measurements. In: Carpenter SR, Kitchell JF (eds) The trophic cascade in lakes. Cambridge University Press. pp. 15–25

Carpenter SR (1993) Statistical analysis of the ecosystem experiments. In: Carpenter SR, Kitchell JF (eds) The trophic cascade in lakes. Cambridge University Press. pp. 26–42
Carpenter SR, Bolgrien D, Lathrop RC, Stow CA, Reed T, Wilson MA (1998) Ecological and economic analysis of lake eutrophication by nonpoint pollution. Austr J Ecol 23:68–79
Carpenter SR, Christensen DL, Cole JJ, Cottingham KL, He X, Hodgson JR, Kitchell JF, Knight SE, Pace ML, Post DM, Schindler DE, Voichick N (1995a) Biological control of eutrophication in lakes. Environ Sci Technol 29:784–786
Carpenter SR, Kitchell JF, Hodgson JR (1985) Cascading trophic interactions and lake productivity. BioScience 35:634–639
Carpenter SR, Kitchell JF, Hodgson JR, Cochran PA, Elser JJ, Elser MM, Lodge DM, Kretchmer D, He X, Von Ende CN (1987) Regulation of lake primary productivity by food web structure. Ecology 68:1863–1876
Carpenter SR, Cottingham KL, Schindler DE (1992) Biotic feedbacks in lake phosphorus cycles. TREE 7:332–336
Carpenter SR, Chrisholm SW, Krebs CJ, Schindler DW, Wright RF (1995b) Ecosystem experiments. Science 269:324–327
Chorus I, Bartram J (eds) (1999) Toxic cyanobacteria in water. A guide to their public health consequences, monitoring and management. E & FN Spon, London
Colbourne JK, Hebert PDN (1996) The systematics of North American Daphnia (Crustacea, Anomopoda): a molecular phylogenetic approach. Phil Trans Roy Soc Lond B 351(1337):349–360
Currie DJ, Dilworth-Christie P, Chapleau F (1999) Assessing the strength of top-down influences on plankton abundance in unmanipulated lakes. Can J Fish Aquat Sci 56(3):427–436.
CUWVO (1976) Schets van de eutrofiëringssituatie van het Nederlandse oppervlaktewater en overzicht van de onderzoeksactiviteiten van de verschillende waterkwaliteitsbeheerders. Resultaten van de eerste eutrofiëringsenquete. Coördinatiecommissie Uitvoering Wet Verontreiniging Oppervlaktewateren, Werkgroep VI
CUWVO (1980) Ontwikkeling van grenswaarden voor doorzicht, chlorophyl, fosfaat en stikstof. Resultaten van de tweede eutrofieringsenquete, november 1980. Coördinatie commissie Uitvoering Wet Verontreiniging Oppervlaktewateren, Werkgroep VI
CUWVO (1988) Samenvatting en conclusie van het vergelijkend onderzoek naar de eutrofiëring in Nederlandse meren en plassen. Resultaten van de derde eutrofiëringsenquête. Coördinatiecommissie Uitvoering Wet Verontreiniging Oppervlaktewateren Werkgroep VI. 26 pp.
Day KE (1989) Acute, chronic and sublethal effects of synthetic pyrethroids on freshwater zooplankton. Environ Toxicol Chem 8:411–416
De Bernardi R, Giussani G (1990) Are blue-green algae a suitable food for zooplankton? An overview. Hydrobiologia 200/201:29–41
De Bruijn J, Crommentuijn T, Van Leeuwen K, Van der Plassche E, Sijm D, Van der Weiden M (1999) Environmental risk limits in the Netherlands. RIVM report No 601640001
DeMott WR (1998) Utilization of a cyanobacterium and a phosphorus-deficient green alga as complementary resources by daphnids. Ecology 79(7):2463–2481
deNoyelles F, Dewey SL, Huggings DG, Kettle WD (1994) Aquatic mesocosms in ecological effects testing: Detecting direct and indirect effects of pesticides. In: Graney RL, Kennedy JH, Rodgers JH (eds) Aquatic mesocosm studies in ecological risk assessment. SETAC Special Publication. Lewis Publishers, Boca Raton pp 577–603
Elser JJ, Urabe J (1999) The stoichiometry of consumer-driven nutrient recycling: theory, observations, and consequences. Ecology 80(3):735–751
Enserink EL (1995) Food mediated life history strategies in Daphnia magna: Their relevance to ecotoxicological evaluations. Thesis Agricultural University Wageningen

Fernández-Casalderrey A, Ferrando MD, Andreu-Moliner E (1993) Effect of the insecticide methylparathion on filtration and ingestion rates of Brachionus calyciflorus and Daphnia magna. Sci Total Environ Suppl:867–876

Fernández-Casalderrey A, Ferrando MD, Andreu-Moliner E (1994) Effect of sublethal concentrations of pesticides on the feeding behavior of Daphnia magna. Ecotoxicol Environ Saf 27:82–89

Fisher TR, Melack JM, Grobbelaar JU, Howarth RW (1995) Nutrient limitation of phytoplankton and eutrophication of inland, estuarine, and marine waters. In: Tiessen H (ed) Phosphorus in the global environment. Scope 54, John Wiley&Sons, Chichester pp. 301–322

Flickinger AL, Bruins RJF, Winner RW, Skillings JH (1982) Filtration and phototactic behavior as indices of chronic copper stress in Daphnia magna Straus. Arch Environ Contam Toxicol 11:457–463

Fliedner A (1997) Ecotoxicity of poorly water-soluble substances. Chemosphere 35:295–305

Foekema EM, Van Dokkum HP (2000) Evaluatie van de blauwalgenbloei in de geestmerambachtplas in 1999. TNO report TNO-MEP R2000/177

Foekema EM, Kaag NHBM, Van Hussel DM, Jak RG, Scholten MCTh, Van de Guchte C (1998) Mesocosm observations on the ecological response of an aquatic community to sediment contamination. Water Sci Technol 37(6–7):249–256

Foekema EM, Jak RG, Scholten MCTh (1996) Toetsing van enkele veldexperimenten in microcosms behandeld met dimethoaat. TNO report TNO-MEP R96/070

Galloway JN (1998) The global nitrogen cycle: changes and consequences. Environ Pollut 102(Suppl 1):15–24

Gliwicz MZ, Sieniawska A (1986) Filtering activity of Daphnia in low concentrations of a pesticide. Limnol Oceanogr 31:1132–1138

Gliwicz ZM (1990) Food thresholds and body size in cladocerans. Nature 343:638–640

Goldman JC (1979) Temperature effects on steady-state growth, phosphorus uptake, and the chemical composition of a marine phytoplankter. Microbial Ecol 5:153–166

Golterman HL (1977) Sediments as a source of phosphate for algal growth. In: HL Golterman (ed), Interactions between sediments and freshwater. Dr W Junk Publishers, The Hague-Pudoc, Wageningen pp. 286–293

Golterman HL (1991) Reflections on post-OECD eutrophication models. Hydrobiologia 218:167–176

Gommers P, Rienks J, Wagemaker FH (1999) Gezuiverde cijfers over zuiveren. RWS RIZA report 99.018

Gosselain V, Viroux L, Descy JP (1998) Can a community of small-bodied grazers control phytoplankton in rivers? Freshwater Biol 39:9–24

Goulden CE, Henry LL, Tessier AJ (1982) Body size, energy reserves, and competitive ability in three species of cladocera. Ecology 63:1780–1789

Grobbelaar JU, House WA (1995) Phosphorus as a limiting resource in inland waters; interactions with nitrogen. In: Tiessen H (ed) Phosphorus in the global environment. Scope 54, John Wiley, Sons, Chichester. pp. 255–273

Grobbelaar JU (1983) Availability to algae of N and P adsorbed on suspended solids in turbid waters of the Amazon river. Arch Hydrobiol 96(3):302–316

Grootelaar EMM, Maas-Diepeveen JL (1988) Invloed van zout- en ammoniagehalte op Daphnia magna (watervlo) en Chironomus riparius (muggelarve). RIZA-report 88/04 7 pp.

Gulati RD, DeMott WR (1997) The role of food quality for zooplankton: remarks on the state-of the art, perspectives and priorities. Freshwater Biol 38:753–768

Hallam TG, Lassiter RR, Li J, McKinney W (1990) Toxicant-induced mortality in models of Daphnia populations. Environ Toxicol Chem 9:597–621

Haney JF, DJ Forsyth, MR James (1994) Inhibition of zooplankton filtering rates by dissolved inhibitors produced by naturally occurring cyanobacteria. Arch Hydrobiol 132:1–13

Hansson LA, LJ Tranvik (1997) Algal species composition and phosphorus recycling at contrasting grazing pressure: an experimental study in sub-Antarctic lakes with two trophic levels. Freshwater Biol 37:45–54

Harper DM, B Brierley, AJD Ferguson, G Phillips (eds) (1999) The ecological bases for lake and reservoir management. Developments in Hydrobiology 136. Kluwer Academic Publishers, Dordrecht

Harrison KE (1990) The role of nutrition in maturation, reproduction and embryonic development of decapod crustaceans: a review. J Shellfish Res 9:1–28

Hartgers EM, EHW Heugens, JW Deneer (1999) Effect of lindane on the clearance rate of Daphnia magna. Arch Environ Contam Toxicol 36(4):399–404

Hatch AC, GA Burton (1999) Sediment toxicity and stormwater runoff in a contaminated receiving system: Consideration of different bioassays in the laboratory and field. Chemosphere 39(6):1001–1017

Havens KE, T Hanazato (1993) Zooplankton community responses to chemical stressors: A comparison of results from acidification and pesticide contamination research. Environ Pollut 82:277–288

Healy FP (1973) Characteristics of phosphorus deficiency in Anabaena. J Phycol 9:383–394

Hecky RE, P Kilham (1988) Nutrient limitation of phytoplankton in freshwater and marine environments: a review of recent evidence on the effects of enrichment. Limnol Oceanogr 33:796–822

Hecky RE, P Campbell, LL Hendzel (1993) The stochiometry of carbon, nitrogen, and phosphorus in particulate matter of lakes and oceans. Limnol Oceanogr 38:709–724

Hessen DO (1997) Stoichiometry in food webs: Lotka revisited. Oikos 79(1):195–200

Hessen DO (1990) Carbon, nitrogen and phosphorus status in Daphnia at varying food conditions. J Plankton Res 12:1239–1249

Hessen DO, E van Donk (1993) Morphological changes in Scenedesmus induced by substances released from Daphnia. Arch Hydrobiol 127:129–140

Hillebrand H, U Sommer (1999) The nutrient stoichiometry of benthic microalgal growth: Redfield proportions are optimal. Limnol Oceanogr 44(2):440–446

Hogenbirk M (1996) De algenproblematiek in het Amstelmeer. Een gebiedsgerichte Ecologische Risico Analyse. Report Department for Ecological Risk Studies DH96/22

Holthaus KIE, HP van Dokkum, DPC van der Veen (2001) Blauwalgenbloei in de Geestmerambachtplas: een evaluatie van de monitoring en maatregelen in 2000. TNO report TNO-MEP R2001/138

Hosper H (1997) Clearing lakes, an ecosystem approach to the restoration and management of shallow lakes in the Netherlands. Thesis Agricultural University Wageningen

Hosper SH, ML Meijer, PA Walker (1992) Handleiding actief biologisch beheer. RIZA/OVB 102 pp.

Hurlbert SH (1975) Secondary effects of pesticides on aquatic ecosystems. Res Rev 58:81–148

Hurlbert SH, MS Mulla, HR Willson (1972) Effects of an organophosphorous insecticide on the phytoplankton, zooplankton, and insect populations of fresh-water ponds. Ecol Monogr 42:269–299

Huwer SL, JM Brils (1999) The role and application of toxicity identification Evaluation (TIE) in the United States. TNO MEP R99/331

Jak R.G. (1997): Toxicant-induced changes in zooplankton communities and consequences for phytoplankton development. Thesis, Free University of Amsterdam. ISBN 90-9010990-0. 125pp

Jak RG, MCTh Scholten (1993) Mesocosm experimenten TNO Den Helder. Indicaties voor het ontstaan van eutrofiëringsverschijnselen. TNO report TNO-MEP R93/026

Jak RG, JL Maas, MCTh Scholten (1996) Evaluation of laboratory derived toxic effect concentrations of a mixture of metals by testing fresh water plankton communities in enclosures. Water Res 30: 1215–1227

Jak RG, JL Maas, MCTh Scholten (1998) Ecotoxicity of 3,4-dichloroaniline in enclosed freshwater plankton communities at different nutrient levels. Ecotoxicology 7:49–60

Jeppesen E, M Sondergaard, JP Jensen, E Mortensen, O Sortkjaer (1996) Fish-induced changes in zooplankton grazing on phytoplankton and bacterioplankton: a long-term study in shallow hypertrophic Lake Sobygaard. J Plankton Res 18:1605–1625

Jeppesen, E, Jensen, JP, Søndergaard, Lauridsen, M, Pedersen, B and Jensen LJ, (1997) Top-down control in freshwater lakes with special emphasis on the role of fish, submerged macrophytes and water depth. Hydrobiologia 342/343: 151–164

Jørgensen SE (1986) Structural dynamic model. Ecol Modell 31:1–9

Kasprzak P, RC Lathrop, SR Carpenter (1999) Influence of different sized Daphnia species on chlorophyll concentration and summer phytoplankton community structure in eutrophic Wisconsin lakes. J Plankton Res 21(11):2161–2174

Kersting K, H van der Honing (1981) Effect of the herbicide dichlobenil on the feeding and filtering rate of Daphnia magna. Verh Internat Verein Limnol 21:1135–1140

Kirk KL, JJ Gilbert (1990) Suspended clay and the population dynamics of planktonic rotifers and cladocerans. Ecology 71:1741–1755

Klein G (1989) Anwendbarkeit des OECD-Vollenweider-Modells auf den Oligotrohierungsprozeß an eutrophierten Gewässern, Vom Wasser 73: 365–373 4394

Knoechel R, B Holtby (1986) Construction and validation of a body-length-based model for the prediction of cladoceran community filtering rates. Limnol Oceanogr 31:1–16

Lampert W, U Sommer (1997) Limnoecology: the ecology of lakes and streams Oxford University Press, Oxford

Lampert W (1987a) Vertical migration of freshwater zooplankton: indirect effects of vertebrate predators on algal communities. In: WC Kerfoot, A Sih (eds), Predation. Direct and indirect impact on aquatic communities. University Press of New England, Hanover pp. 291–299

Lampert W (1987b) Laboratory studies on zooplankton-cyanobacteria interactions. NZ J Mar Freshwater Res 21:483–490

Lampert W (1987c) Feeding and nutrition in Daphnia. In: RH Peters, R De Bernardi (eds), Daphnia. Mem Ist Ital Idrobiol 45:143–192

Lampert W (1988) The relationship between zooplankton biomass and grazing: A review. Limnologica 19:11–20

Lampert W, W Fleckner, H Rai, BE Taylor (1986) Phytoplankton control by grazing zooplankton: a study on the spring clear-water phase. Limnol Oceanogr 31:478–490

Lathrop RC, SR Carpenter, LG Rudstam (1996) Water clarity in lake Mendota since 1900: responses to differing levels of nutrients and herbivory. Can J Fish Aquat Sci 53:2250–2261

Lauridsen TL, DM Lodge (1996) Avoidance by Daphnia magna of fish and macrophytes: chemical cues and predator-mediated use of macrophyte habitat. Limnol Oceanogr 41:794–798

Lehman JT (1991) Interacting growth and loss rates: The balance of top-down and bottom-up controls in plankton communities. Limnol Oceanogr 36:1546–1554

Leibold M, HM Wilbur (1992) Interactions between food-web structure and nutrients on pond organisms, Nature 360:341–343

Likens GE (1972) Eutrophication and aquatic ecosystems. In: GE Likens (ed), Nutrients and eutrophication: the limiting nutrient controversy. Limnol Oceanogr Special Symposia 13–13

Lürling M, E van Donk (1997) Life history consequences for Daphnia pulex feeding on nutrient limited phytoplankton. Freshwater Biol 38:693–710

Lyche A, T Andersen, K Christoffersen, DO Hessen, PH Berger Hansen, A Klysner (1996) Mesocosm tracer studies. 1 Zooplankton as sources and sinks in the pelagic phosphorus cycle of a mesotrophic lake. Limnol Oceanogr 41:460–474

Lynch M (1980) The evolution of cladoceran life histories. Quart Rev Biol 55:23–42

Mark U, J Solbé (1998) Analysis of the ECETOC Aquatic Toxicity (EAT) database. V-The relevance of Daphnia magna as a representative test species. Chemosphere 36:155–166

Matveev V, L Matveeva (1997) Grazer control and nutrient limitation of phytoplankton biomass in two Australian reservoirs. Freshwater Biol 38:49–65

Matveev V, W Gabriel (1994) Competitive exclusion in cladocera through elevated mortality of adults. J Plankton Res 16:1083–1094

Matveev V, L Matveeva, GJ Jones (1994) Study of the ability of Daphnia carinata King to control phytoplankton and resist cyanobacterial toxicity: implications for biomanipulation in Australia. Austr J Mar Freshwater Res 45(5):889–904

Mazumder A, DRS Lean (1994) Consumer-dependent responses of lake ecosystems to nutrient loading. J Plankton Res 16:1567–1580

Mazumder A (1993) Phosphorus-chlorophyll relationships under contrasting zooplankton community structure: Potential mechanisms. Can J Fish Aquat Sci 51:401–407

McCauley E, WW Murdoch, S Watson (1988) Simple models and variation in plankton densities among lakes. Am Nat 132:383–403

McMahon JW, FH Rigler (1965) Feeding rate of Daphnia magna Straus in different foods labeled with radioactive phosphorus. Limnol Oceanogr 10:105–113

McQueen DJ (1998) Freshwater food web biomanipulation: a powerful tool for water quality improvement, but maintenance is required. Lakes Reserv Res Managem 3:83–94

Meador JP (1991) The interaction of pH, dissolved organic carbon, and total copper in the determination of ionic copper and toxicity. Aquat Toxicol 19:13–32

Meijer ML (2000) Biomanipulation in the Netherlands, 15 years of experience. Thesis Agricultural University Wageningen

Mjelde M, BA Faafeng (1997) Ceratophyllum demersum hampers phytoplankton development in some small Norwegian lakes over a wide range of phosphorus concentrations and geographical latitude. Freshwater Biol 37:355–365

Moss B, H Balls (1989) Phytoplankton distribution in a floodplain lake and river system. II Seasonal changes in the phytoplankton communities and their control by hydrology and nutrient availability. J Plankton Res 11:839–867

Moss B (1998) Shallow lakes biomanipulation and eutrophication. Scope Newsletter 29

Moss B, J Madgwick, G Phillips (1996a) A guide to the restoration of nutrient-enriched shallow lakes Broads Authority, Norwich, UK

Moss B, J Stansfield, K Irvine (1991) Development of daphnid communities in diatom- and cyanophyte-dominated lakes and their relevance to lake restoration by manipulation. J Appl Ecol 28:586–602

Moss B, J Stansfield, K Irvine, M Perrows, G Phillips (1996b) Progressive restoration of a shallow lake: A 12-year experiment in isolation, sediment removal and biomanipulation. J Appl Ecol 33:71–86

Moss B, S McGowan, L Carvalho (1994) Determination of phytoplankton crops by top-down and bottom-up mechanisms in a group of English lakes, the West Midland meres. Limnol Oceanogr 39:1020–1029

Mount DR, DD Gulley, JR Hockett, TD Garrison, JM Evans (1997) Statistical models to predict the toxicity of major ions to Ceriodaphnia dubia, Daphnia magna and Pimephales promelas (Fathead minnows). Environ Toxicol Chem 16:2009–2019

Muñoz MJ, C Ramos, JV Tarazona (1996) Bioaccumulation and toxicity of hexachlorobenzene in Chlorella vulgaris and Daphnia magna. Aquat Toxicol 35:211–220

Odum EP (1985) Trends expected in stressed ecosystems. BioScience 35:419–422

OECD (1982) Eutrophication of waters. Monitoring, assessment and control. OECD, Paris

OECD (1993a) OECD guidelines for the testing of chemicals. Organisation for Economic Co-operation and Development, Paris 2 vol.

OECD (1993b) Guideline for testing of chemicals no 201: "Algal Growth inhibition test" adopted 7 June 1984, Organisation for economic Co-operation and Development, Paris (1993)

Ogilvie BG, SF Mitchell (1998) Does sediment resuspension have persistent effects on phytoplankton? Experimental studies in three shallow lakes. Freshwater Biol 40:51–64

Ordelman HGK, PBM Stortelder, TEM ten Hulscher, FH Wagemaker, JM van Steenwijk, J Botterweg, PCM Frintrop, HG Evers (1993a) Watersysteemverkenningen. Carbamaten. RIZA report 93.010 / DGW report 93.022

Ordelman HGK, PCM van Noort, JM van Steenwijk, TEM ten Hulscher, HG Evers (1993c) Watersysteemverkenningen. Triazinen. RWS RIZA report 93.036/ DGW RIKZ report 93.050 182 pp.

Ordelman HGK, PCM van Noort, JM van Steenwijk, TEM ten Hulscher, MA Beek, J Botterweg, R Faasen, PCM Frintrop, HG Evers (1993b) Watersysteemverkenningen. Dithiocarbamaten. RWS RIZA report 93.025/ DGW report 93.041 154 pp.

Ordelman HGK, PCM van Noort, TEM ten Hulscher, MA Beek, JM van Steenwijk, PCM Frintrop, EHG Evers (1994) Watersysteemverkenningen. Organofosforbestrijdingsmiddelen. RWS RIZA nota 94.043/ DGW report RIKZ-94.028 186 pp. Plus annexes

Ostroumov AP (2002) Inhibitory analysis of top-down control: new keys to studying eutrophication, algal blooms, and self-purification. Hydrobiologia 496:117–129

Overbeck J (1989) Qualitative and quantitative assessment of the problem. In: SE Jörgensen, RA Vollenweider (eds), Guidelines of lake management. Vol 1 Principles of lake management. UNEP International Lake Environment Committee pp. 43–52

Parsons T (1982) The future of controlled ecosystem enclosure experiments. In: GD Grice, MR Reeve (eds), Marine mosocosms. Biological and chemical research in experimental ecosystems. Springer-Verlag, New York pp. 411–418

Pastor J, D Binkley (1998) Nitrogen fixation and the mass balances of carbon and nitrogen in ecosystems. Biogeochemistry 43(1):63–78

Persoone G, CR Janssen (1993) Freshwater invertebrate toxicity tests. In: P Calow (ed), Handbook of ecotoxicology. Vol. 1 Blackwell Scientific Publications, Oxford pp. 51–65

Portielje R, DT van der Molen (1997a) Trendanalyse eutrofiëringstoestand van de Nederlandse meren en plassen. Deelrapport I voor de Vierde Eutrofiëringsenquete RWS RIZA report 97.060

Portielje R, DT van der Molen (1997b) Relaties tussen eutrofiëringsvariabelen en systeemkenmerken van de Nederlandse meren en plassen, RIZA report 98.007

Portielje R, DT van der Molen (1998) Trend-analysis of eutrophication variables in lakes in The Netherlands. Water Sci Technol 37(3):235–240

Pott E (1980) Die Hemmung der Futteraufnahme von Daphnia pulex – eine neue limnotoxikologische Messgrösse. Z Wasser Abwasser Forsch 13(2):52–54

Reading JT, AL Buikema (1980) Effects of sublethal concentrations of Selenium on metabolism and filtering rate of Daphnia pulex. Bull Environ Contam Toxicol 24:929–935

Reckendorfer W Keckeis H Winkler G Schiemer F (1999) Zooplankton abundance in the River Danube, Austria: the significance of inshore retention. Freshwater Biol 41(3):583–591

Repka S (1996) Inter- and intraspecific differences in Daphnia life histories in response to two food sources: the green alga Scenedesmus and filamentous cyanobacterium Oscillatoria. J Plankton Res 18:1213–1223

Repka S (1997) Effects of food type on the life history of Daphnia clones from lakes differing in trophic state. I Daphnia galatea feeding on Scenedesmus and Oscillatoria. Freshwater Biol 38:675–683

Reynolds CS (1989) Physical determinants of phytoplankton succession. In: U Sommer (ed), Plankton ecology. Springer-Verlag, Berlin pp. 9–56

Reynolds CS (1992) Eutrophication and the management of planktonic algae: What Vollenweider couldn't tell us. In: DW Sutcliffe, JG Jones (eds), Eutrophication: Research and application to water supply. Freshwater Biological Association Special Publication, Ambleside pp. 4–29

Reynolds CS (1994) The ecological basis for the successful biomanipulation of aquatic communities. Arch Hydrobiol 130:1–33

RIVM (1995) Achtergronden bij: Milieubalans 95. Samsom HD Tjeenk Willink, Alphen aan den Rijn

Rohrlack T, M Henning, JG Kohl (1999) Mechanisms of the inhibitory effect of the cyanobacterium Microcystis aeruginosa on Daphnia galeata's ingestion rate. J Plankton Res 21(8):1489–1500

Romanovsky YE (1984) Individual growth rate as a measure of competitive advantages in cladoceran crustaceans. Int Rev Ges Hydrobiol 69:613–632

Romanovsky YE (1985) Food limitation and life-history strategies in cladoceran crustaceans. Arch Hydrobiol Beih Ergebn Limnol 21:363–372

Rothaupt KO (1997) Grazing and nutrient influences of Daphnia and Eudiaptomus on phytoplankton in laboratory microcosms. J Plankton Res 19:125–139

Salmaso N, F Decet, P Cordella (1999) Understanding deep oligotrophic subalpine lakes for efficient management. Hydrobiologia 396:253–263

Sarnelle O (1992) Nutrient enrichment and grazer effects on phytoplankton in lakes. Ecology 73:551–560

Scheffer M (1998) Ecology of shallow lakes. Population and Community Biology Series 22. Chapman, Hall, London

Scheffer M, SH Hosper, M-L Meijer, B Moss, E Jeppesen (1993) Alternative equilibria in shallow lakes. TREE 8:275–279

Schreurs H (1992) Cyanobacterial dominance. Relations to eutrophication and lake morphology. Thesis University of Amsterdam

Sharpley AN, S Rekolainen (1997): Phosphorus in agriculture and its environmental implications. In: H Tunney, OT Carton, PC Brookes, AE Johnston (eds), Phosphorus loss from soil to water. CAB International, New York pp. 1–53

Sharpley AN, MJ Hedley, E Sibbesen, A Hillbricht-Ilkowska, WA House, L Ryszkowski (1995) Phosphorus transfers from terrestrial to aquatic ecosystems. In: H Tiessen (ed), Phosphorus in the global environment. Scope 54, John Wiley, Sons, Chichester pp. 177–199

Sommer U (1989) The role of competition for resources in phytoplankton succession. In: U Sommer (ed), Plankton ecology. Springer-Verlag, Berlin pp. 57–106

Sommer U (1992) Phosphorus-limited Daphnia: intraspecific facilitation instead of competition. Limnol Oceanogr 37:966–973

Sommer U, ZM Gliwicz, W Lampert, A Duncan (1986) The PEG-model of seasonal succession of planktonic events in fresh waters. Arch Hydrobiol 106:433–471

Spencer CN, BK Ellis (1998) Role of nutrients and zooplankton in regulation of phytoplankton in Flathead Lake (Montana, USA), a large oligotrophic lake. Freshwater Biol 39:755–764

Stansfield J, B Moss, K Irvine (1989) The loss of submerged plants with eutrophication. III Potential role of organochlorine pesticides: a paleooecological study. Freshwater Biol 22:109–132

Stephen D, B Moss, G Phillips (1998) The relative importance of top-down and bottom-up control of phytoplankton in a shallow macrophyte-dominated lake. Freshwater Biol 39:699–714

Sterner RW (1993) Daphnia growth on varying quality of Scenedesmus: Mineral limitation of zooplankton. Ecology 74:2351–2360

Streble H, D Krauter (1988) Leben im Wassertropfen. Mikroflora und Mikrofauna des Süsswassers. Franckh-Kosmos Verlags, Stuttgart

Stumm W, JJ Morgan (1996) Aquatic chemistry. Chemical equilibria and rates in natural waters. John Wiley, Sons, New York 3rd ed

Suedel BC, E Deaver, JH Rodgers (1996) Formulated sediment as a reference and dilution sediment in definitive toxicity tests. Arch Environ Contam Toxicol 30:47–52

Taylor G, DJ Baird, AMVM Soares (1998) Surface binding of contaminants by algae: consequences for lethal toxicity and feeding to Daphnia magna Straus. Environ Toxicol Chem 17:412–419

Teunissen-Ordelman HGK, SM Schrap (1996b) Watersysteemverkenningen. Bestrijdingsmiddelen. RWS RIZA report 96.040

Teunissen-Ordelman HGK, PCM van Noort, JM van Steenwijk, MA Beek, ThEM ten Hulscher, PCM Frintrop, R Faasen (1995b) Watersysteemverkenningen. Bentazon en chloridazon. RWS RIZA report 95.046

Teunissen-Ordelman HGK, PCM van Noort, MA Beek, TEM ten Hulscher, JM van Steenwijk, R Faasen, PCM Frintrop (1995a) Watersysteemverkenningen. Chloorfenoxycarbonzuren. RWS RIZA report 95.007

Teunissen-Ordelman HGK, PCM van Noort, SM Schrap, MA Beek, JM van Steenwijk, R Faasen, PCM Frintrop (1995c) Watersysteemverkenningen. Nitroanilinen. RWS RIZA draft report

Teunissen-Ordelman HGK, SM Schrap, PCM van Noort, JM van Steenwijk, MA Beek, R Faasen, PCM Frintrop (1996c) Watersysteemverkenningen. Aniliden. RWS RIZA report 97.021

Teunissen-Ordelman HGK, SM Schrap, PCM van Noort, MA Beek, R Faasen, JM van Steenwijk, PCM Frintrop (1996a) Watersysteemverkenningen. Pyrethroïden. RWS RIZA draft report

Thompson JM, AJD Ferguson, CS Reynolds (1982) Natural filtration rates of zooplankton in a closed system: The derivation of a community grazing index. J Plankton Res 4:545–560

Tilman D, SS Kilham, P Kilham (1982) Phytoplankton community ecology: The role of limiting nutrients. Ann Rev Ecol Syst 13:349–372

Timms RM, B Moss (1984) Prevention of growth of potentially dense phytoplankton populations by zooplankton grazing, in the presence of zooplanktivorous fish, in a shallow wetland ecosystem. Limnol Oceanogr 29:472–486

Torres-Orozco R E, SA Zanatta (1998) Species composition, abundance and distribution of zooplankton in a tropical eutrophic lake: Lake Catemaco, México. Tropical Biol 46(2)

Urabe J, M Nakanishi, K Kawabata (1995) Contribution of metazoan plankton to the cycling of nitrogen and phosphorus in Lake Biwa. Limnol Oceanogr 40:232–241

Valiela I (1993) Ecology of water columns. In: RSK Barnes, KH Mann (eds), Fundamentals of aquatic ecology. Blackwell Scientific Publications, Oxford 2nd ed pp. 29–56

Van der Gun JHJ, J Joziasse (1999) In-situ waterbodemsanering: voorstelbaar en haalbaar? Rapporten Programma Geïntegreerd Bodemonderzoek deel 23, ISBN 90-73270-38-3

Van der Helm R (2000) Rapportage signalerend onderzoek bestrijdingsmiddelen in hoofdwatersysteem Hollands Noorderkwartier, 1998. Hoogheemraadschap van Uitwaterende Sluizen in Hollands Noorderkwartier

Van Dokkum HP, G Hoornsman (2000) Monitoring blauwalgen in de Geestmerambachtplas in de zomer van 1999. TNO report TNO-MEP -R 2000/007

Van Dokkum HP, G Hoornsman, DPC van der Veen (1999) Blauwalgenbloei in de Geestmerambachtplas in de zomer van 1998. TNO rapport TNO report TNO-MEP – R99/047

Van Donk E, DO Hessen (1993) Grazing resistance in nutrient-stressed phytoplankton. Oecologia 93:508–511

Van Donk E, M Lürling, DO Hessen, GM Lokhorst (1997) Altered cell wall morphology in nutrient-deficient phytoplankton and its impact on grazers. Limnol Oceanogr 42:357–364

Van Luijn F, DT van der Molen, WJ Luttmer, PCM Boers (1995) Influence of benthic diatoms on the nutrient release from sediments of shallow lakes recovering from eutrophication. Water Sci Technol 32:89–97

Van Steenwijk JM, GAJ Mol (1996) Toetsing huidige en verwachte water(bodem)kwaliteit aan de grenswaarden. RWS RIZA notanr 95.063

Vanni MJ, CD Layne (1997) Nutrient recycling and herbivory as mechanisms in the "top-down" effect of fish on algae in lakes. Ecology 78:21–40

Vanni MJ, C Luecke, JF Kitchell, JJ Magnuson (1990) Effects of planktivorous fish mass mortality on the plankton community of Lake Mendota, Wisconsin: implications for biomanipulation. Hydrobiologia 200/201:329–336

Vijverberg, J, Gulati, RD, Mooij, WM, (1993) Food-web studies in shallow eutrophic lakes by the Netherlands institute of ecology: main results, knowledge gaps and new perspectives, 27(1): 35–49 4128

Vitousek PM, JD Aber, RW Howarth, GE Likens, PA Matson, DW Schindler, WH Schlesinger, DG Tilman (1997) Human alteration of the global nitrogen cycle: sources and consequences. Ecol Appl 7:737–750

Vollenweider RA, J Kerekes (1980) The loading concept as basis for controlling eutrophication philosophy and preliminary results of the OECD programme on eutrophication. Prog Water Technol 12:5–38

Vollenweider RA (1990) Eutrophication: Conventional and non-conventional considerations and comments on selected topics. In: R de Bernardi, G Giussani, L Barbanti (eds), Scientific perspectives in theoretical and applied limnology. Mem Ist Ital Idrobiol 47:77–134

Von Sperling E (1997) The process of biomass formation as the key point in the restoration of tropical eutrophic lakes. Hydrobiologia 342/343:351–354

Wang W (1987) Factors affecting metal toxicity to (and accumulation by) aquatic organisms – overview. Environ Int 13:437–457

Wetzel RG (1990) Detritus, macrophytes and nutrient cycling in lakes in Scientific perspectives in theoretical and applied limnology. In De Bernadi R, G Giussani and L Barbanti (eds). Mem Ist Ital Idrobiol, 47:233–249

Winner RW, HA Owen, MV Moore (1990) Seasonal variability in the sensitivity of freshwater lenthic communities to a chronic copper stress. Aquat Toxicol 17:75–92

WL (1996) Bestrijding van eutrofiëring in de Geestmerambachtplas. WL rapport T1245

Yasuno M, A Asaka, Y Kono (1993) Effects of pyraclofos (an organophosphorous insecticide) on nutrient enriched ecosystems. Chemosphere 27:1813–1824

Zaffagnini F (1987) Reproduction in Daphnia. In: RH Peters, R De Bernardi (eds), Daphnia. Mem Ist Ital Idrobiol 45:245–284

Zurek R, H Bucka (1994) Algal size classes and phytoplankton-zooplankton interacting effects. J Plankton Res 16:583–601